5/01	13	04/2000
1/04	17	12/03
9/05	18-1	2/05
7/07	20-1	5/06
8/09	21-1	8/07
4/2012	22-1	12/2011
		AP '94

Women with Wings

FEMALE FLYERS IN FACT AND FICTION

ALSO BY MARY CADOGAN

Richmal Crompton: The Woman Behind William
Frank Richards: The Chap Behind the Chums
The William Companion

WITH PATRICIA CRAIG
You're a Brick, Angela!
Women and Children First
The Lady Investigates

MARY CADOGAN

Women with Wings

FEMALE FLYERS IN FACT AND FICTION

ACADEMY CHICAGO PUBLISHERS

Published in 1993 by
Academy Chicago Publishers
213 West Institute Place
Chicago, Illinois 60610

Printed and bound in the U.S.A.
Printed on acid-free paper.

Library of Congress Cataloging-in-Publication Data

Cadogan, Mary.
    Women with wings : female flyers in fact and fiction / Mary
Cadogan.
        p.   cm.
    Includes bibliographical references and index.
    ISBN 0-89733-385-3
    1. Women in aeronautics. 2. Women in literature. I. Title.
TL544.C23 1992
629. 13' 0082—dc20   92-46541
                                    CIP

629.13

For my daughter, Teresa Mary:
and in memory of the great aviatrices of the past,
especially Pauline Gower who, as well as inspiration,
has provided the title, *Women with Wings*,
which was first used for her 1938 autobiographical book.

Picture acknowledgements

Chapter Illustrations

Title page: *A Tribute to Women Aviators*, painting by Roderick Lovesay 1991. Chapter 4: Express Newspapers. Chapter 10: Academy Chicago Publishers. Chapter 11: © Fleetway Editions Ltd, from *Girl* Magazine. Chapter 12: Mills & Boon Ltd. Chapter 13: D. C. Thomson & Co Ltd.

Photo Section (between pages 156-157)

2. Courtesy of Molly Sedgwick. 3. Courtesy of Molly Sedgwick. 4. *Flight* magazine. 6. John W. Underwood. 11. Express Newspapers. 14. Arcadia Publications. 18. John Fairfax Group PTY Ltd. 19. New Zealand Herald. 21. The *Illustrated London News* Picture Library. 22. The Royal Aeronautical Society. 26. Academy Chicago Publishers. 29. National Air and Space Museum, Washington. 31. © Fleetway Editions Ltd (from *Girl* magazine, 1958). 32. © Fleetway Editions Ltd, *Eagle* magazine. 33. © Fleetway Editions Ltd, *Eagle* Magazine. 34. National Periodical Publications Inc. 35. TM and © 1992 Cynthy J. Wood and Innovative Corp; all rights reserved; art by Bart Sears and Bill Anderson. 36. Royal Aeronautical Society. 37. Sygma (M. Polak). 38. Science Photo Library (Novosti). 39. J.P. Laffont. 40. Popperfoto London. 42. Eastbourne *Herald*.

CONTENTS

ACKNOWLEDGEMENTS

In writing *Women with Wings* I have received generous help with research information and the finding of elusive books of reference. I owe particular thanks to Squadron Leader Dennis L. Bird, RAF (retired) who has given me invaluable technical advice as well as enthusiastic encouragement. I would like to thank Flight Lieutenant Julie Gibson and Flying Officer Anne-Marie Dawe who have allowed themselves to be interviewed for this book; also, for contributing expertise, information or research material, Jennifer Schofield (Piers Williams), Norman Wright, Denis Gifford, Roger Coombes, Robert Adey, Jack Adrian, Richard Hollis, Margaret Heaton, David Pringle, the Press Office of the Turkish Embassy in London, and my husband, Alexander Cadogan. I am grateful to my editors at Macmillan, Susanna Wadeson, Jane Wood and Hazel Orme. I also owe thanks to editors Anita and Jordan Miller, and to production staff Julia Anderson-Miller and Barbara Spann at Academy Chicago Publishers.

MARY CADOGAN

INTRODUCTION

Go forward therefore with this your new and admirable skill, by which you do thus climb up to the stars!

SIR THOMAS MORE
in a letter to his children

THIS book is a celebration of the lives and achievements of women who have taken to the skies, from the early balloonists to contemporary astronauts. As parachutists, wing-walkers, stunt flyers, record-makers and -breakers, stewardesses, civil and military pilots, as well as in back-up roles on the ground, such women have constantly defied prejudices and proved their skill, endurance and integrity.

Their participation in aviation is part of the larger story of society's struggle to adapt to changing technologies and attitudes, which involves us all. Flying implies liberation and control of the elements. Female aviators, by their exploits and in their own personalities, have become particularly symbolic of the widening spheres of activity sought by women and girls. Until comparatively recently, however, even those who were stirringly acclaimed as heroines by writers, film-makers, politicians, royalty and the general public had difficulty in finding sustained and responsible work in their fields of expertise. (For example, America's first woman flyer, Blanche Stuart Scott, abandoned aviation in 1916 because it offered no place for female pilots or mechanics, while in England, two decades later, Amy Johnson experienced similar frustrations.)

Every decade of the twentieth century has produced its own particular aeronautical luminaries. The parachutist Dolly Shepherd captured the imagination of the Edwardians; the

3

1920s were the years of stunt women in America and aristocratic lady flyers in England; the 1930s proved to be the heroic decade, when Amelia Earhart, Amy Johnson, Jean Batten, Beryl Markham and others became Queens of the Air; in the 1940s women on both sides of the Atlantic played key wartime roles by ferrying planes from factories to air-fields, while Hanna Reitsch test-piloted in Nazi Germany and, in Soviet Russia, three regiments of women flew fighters and bombers into combat. In the 1950s, Jacqueline Cochran, the first woman to break the sound barrier, competed regu-larly with test pilot Jacqueline Auriol for world speed re-cords. Sheila Scott harnessed modern technology to perform remarkable feats of long-distance aviation in the 1960s, which also marked the sending of the first female astronaut, Valentina Tereshkova, into space. The 1970s and 1980s saw further long-distance endurance records by individuals, no-tably Judith Chisholm and Jeana Yeager, and also the cor-porate campaigns of women for equal rights and equal responsibilities, which culminated in their employment as pilots of civil airliners and military jets.

The achievements of real-life flyers have been colourfully reflected, encouraged, frowned upon and fantasised in popular fiction, from comics, story-papers and novels for girls to adult romances from Mills & Boon and other pub-lishers. Generally speaking, fictional aviatrices are intrepid, fetching and fearfully charismatic. Career success comes more easily to them than to their factual counterparts, and their feminism is of the exuberant and simplistic variety. ('Who started the war, anyway? Men. Take a look at the world and see what a nice mess men have made of it. No wonder they had to appeal to women to help them out,' snorts W. E. Johns's teenage air ace, Worrals, in 1945.)

The interrelationships between fact and fantasy in female aviation make intriguing reading. It is particularly interest-ing to see how differently aviatrices are treated in girls' weeklies (where they are exciting symbols of liberation) and in certain stories and comic-strips for men. The latter show them as victims or vampires, and potent metaphors of sexual energy and enticement such as Wonderwoman, who has

4

survived resiliently over five decades. It is worthy of note that, despite the influence of feminism, the less progressive attributes of Wonderwoman are expressed and extended today far more luridly than they were during the 1940s when she was first launched. Transported into high-tech settings and embodied in characters like the Stargrazers ('sexy sailors of the space age', see chapter 13), such fictional aviatrices might seem little more than sex symbols.

Counterbalancing this trend, several recent novels with both contemporary and retrospective backgrounds deal sympathetically and authentically with the characterisation and career ambitions of female flyers. Ten years after the beginning of the twentieth century women and girls who took to the air began to be featured in fiction. In the century's final decade they remain fascinating expressions of frequently conflicting views about feminine roles in aviation – and in society generally. Hopefully, by the beginning of the twenty-first century, women with wings will be presented in fiction with the perception and seriousness that the achievements of their real-life counterparts so well deserve.

MARY CADOGAN
Beckenham, 1991

CHAPTER ONE

Soaring and 'Chuting

Mrs Letitia Sage, the first British female 'aeronaut', making a
balloon ascent in 1785.

As soon as the balloon was set free, the silence and the tension were broken by a roar from the spectators . . . our passengers responded by waving their handkerchiefs from within the basket. I did not join in all this waving. I was too busy holding on.

DOLLY SHEPHERD
When the 'Chute Went Up (1984)

I T WAS not until the eighteenth century that would-be aeronauts began to experiment seriously with the balloon as a vehicle for bearing themselves aloft. Early pioneers of flight had been obsessed with the concept of the 'ornithopter' or bird-machine. Those who vainly flapped waxen, feathered, wooden or fabric wings before falling to their deaths included Icarus in classical Greek legend, and, in real life, the Anglo-Saxon King Bladud and a variety of courageous or headstrong experimenters who hurled themselves from the tops of cliffs and towers during the Middle Ages.

Fortunately the multi-faceted Leonardo da Vinci was never rash enough to build and attempt to fly the beautifully drawn ornithopters (or the helicopter and pyramidal-tent parachute) which he designed in the fifteenth century. His posthumously published writings indicate that he eventually realised the limitations of wing-waggling methods for human flight, and in 1680 Borelli, another Italian, produced a book which convinced most potential aviators of the futility of trying to emulate bird flight methods. At about the same time the Brazilian priest Bartolomeo de Gusmao made a model hot-air balloon and an early type of parachute to help it off the ground and control its descent. There is no record of his balloon ever making an outdoor flight, but developments in ballooning and parachuting thereafter seemed to proceed almost in parallel.

It was in France that the first successful balloon ascent was made. After several years of experimenting with paper bags filled with hot air, the brothers Joseph and Etienne Montgolfier sent up a passenger-carrying balloon on 19 September 1783 at Versailles, watched by a crowd of 30,000. The event was graced by the presence of King Louis XVI and Queen Marie-Antoinette, but the passengers slung in a basket beneath the fragile craft were less distinguished: they were a sheep, a duck and a cockerel. After eight minutes their balloon came down and crashed on to a tree. Fortunately none of the animals appeared to have been seriously hurt. Despite their survival, the king felt apprehensive about the hazards of this new form of locomotion. He insisted that flights scheduled to carry human passengers should use only convicted criminals. However, one of the Montgolfier brothers' assistants was soon permitted to make a tethered ascent; the Marquis d'Arlandes then became the first person ever to make a free flight, and other aeronauts started to go up in balloons over France, Italy, England and elsewhere.

On 20 May 1784, not long after the menagerie and various intrepid male balloonists had soared aloft, the first feminine ascent took place. A group of ladies sat in the basket of a tethered Montgolfier hot-air balloon which floated impressively over Paris from the Faubourg St-Antoine. Three of them – the Marquise, the Comtesse de Montalembert and the Comtesse de Podenas – were titled; Mademoiselle de Lagarde completed the company. There is no record of any of them participating in further aeronautical exploits, but a couple of weeks after their ascent a *compatriote*, Madame Thible, became the first woman to be carried in a free-flight balloon. She took off from Lyon with a Monsieur Fleurant on 4 June 1784. They reached an altitude of 8500 feet in a 'Montgolfière' named *Le Gustav*, presumably in honour of the King of Sweden who watched the procedure from the ground. Contemporary records emphasise Madame Thible's

stylishness as well as her daring. Although a far cry from the
more practical, yet still engagingly colourful, garb which
female balloonists were to adopt soon afterwards, her lace-
trimmed dress and feathered hat were extremely fetching.
Madame Thible expressed her appreciation of being hoisted
into the skies with some hearty and prolonged singing but,
like the four ladies who preceded her, she was never tempted
to make another ascent.

Just about a year after the Montgolfier brothers' pioneering flight, Vincent Lunardi, who was on the staff of the
Neapolitan ambassador to Britain, made a much-publicised,
lengthy balloon voyage from London to Ware. Jean-Pierre
Blanchard had already established a reputation for aerial
showmanship in France. He came to England in the autumn
of 1784 and started to compete with Lunardi for public
acclaim.

One way of publicising themselves and their activities, of
course, was to carry interesting and celebrated passengers.
Blanchard decided that taking up a young female would not
only create a lot of attention but dispel people's doubts about
the safety of ballooning. In May 1785 he persuaded a
fourteen-year-old French girl, Mademoiselle Simonet, to go
up with him. However, once the balloon had reached the
height of 30 feet, her hysterically fearful shrieks forced him
down to earth again. In spite of this ignominious descent,
Blanchard's reluctant passenger made her mark in history as
the first member of her sex to float, if only briefly, over
England.

Not to be outdone, in the following month Lunardi
arranged to go up from St George's Fields, London, with
Colonel Hastings, George Biggin and Mrs Letitia Ann Sage,
a handsome but not particularly outstanding actress. She
thus became the first British woman aeronaut. According to
a chronicler of the flight, Mrs Sage had been chosen from
several actresses who applied. Lunardi was seeking both
'beauty and ballast'; Letitia certainly provided the latter, and
the showman got rather more than he bargained for. His
female passenger's excessive weight led to Lunardi having to
step out of the basket with Colonel Hastings, leaving the

journey to Biggin and Letitia who managed, despite their inexperience, to rise and descend without mishap. They came to earth in Harrow where boys from the famous public school rescued them from the wrath of a farmer in whose field they had landed. Letitia celebrated by having an excellent dinner locally.

Soon afterwards she wrote 'A Letter Describing the General Appearances and Effects of the Expedition with Lunardi's Balloon' in which she commented on being 'infinitely better pleased than ever I was with any former event of my life'. There is no doubt that this aerial exploit was felicitous for her: apparently she subsequently never had difficulty in finding congenial employment, although her taste for adventure, like that of those earlier female balloonists, was completely assuaged by a once-only ascent.

The actress's taking to the air seems to have been one up for Lunardi in his rivalry with Blanchard, but the French enthusiast indefatigably continued his experiments. A serious problem for pioneer aeronauts was the fact that neither the hot-air nor the hydrogen balloon was directable. Dependence on wind made the direction and length of flights, and the location of landings, a chancy business. In 1795 Blanchard produced a form of parachute to help to steer a balloon, which he then used to drop a dog in a wicker basket. Pleased with the results of these experiments, he continued to work with parachutes as aids to dirigibility, and toured Europe and America giving demonstrations in which he subjected cats and dogs to parachute descents.

His disastrous experience with the noisily protesting Mademoiselle Simonet might well have put him off the idea of ballooning with females, but, undeterred, he made several ascents with his wife, Madeleine Sophie Blanchard. She was said to be 'small, ugly and nervous': however, her anxieties may have embraced the hazards of earth, but not those of the air. She was frightened of being enclosed in a horse-drawn carriage, and hated many kinds of noise, but found escape and peace when borne aloft in a balloon. Sadly, her life seems to have been littered with ironical disasters: her husband died in 1809 after a heart attack suffered during a ballooning trip; Sophie, who gamely continued his good and intrepid works on her own, herself became the first woman to die in an aerial accident. Her hydrogen balloon was ignited when she was giving a firework display over the Tivoli Gardens in Paris in 1819. Once the hydrogen was alight the balloon rapidly lost height and hit a roof-top: Sophie fell out and broke her neck. If she and her husband had pursued their parachuting experiments further and devised one able to break the fall of a man or woman, she might well have escaped this disaster. Her reputation is not confined to hindsight and *The Guinness Book of Aircraft Facts and Feats*: Napoleon had made her Official Aeronaut of the Empire during her lifetime. She was not only the first woman to die in the air, but the first to ascend in a balloon in her own right (which she did in 1805) rather than as a passenger.

The French remained in the forefront of aviation. André-Jacques Garnerin designed the first parachute capable of taking the weight of a human body in 1797, but this was not adopted generally until it had undergone several improvements suggested by him and others. In 1815 his wife became the first female to parachute to earth. She never repeated her drop, but Eliza Garnerin, their niece, was to give many parachuting demonstrations between 1816 and 1836. She became the first professional woman parachutist, whose jumping seems to have drawn large and regular crowds.

A British female aeronaut was then to achieve brief but bizarre celebrity. Mrs Graham, who has since been described as 'a combination of suffragette and showgirl', ascended in August 1836 with the Duke of Brunswick, which apparently prompted some mean-spirited speculation about whether or not immorality was possible in the basket of a balloon. Gossips had little to get their teeth into when Mrs Graham next went up a year or two later – this time accompanied by an all-female party – in the Victoria and Albert Balloon. Tongues wagged again, however, when on another trip she fell from a balloon at a height of 100 feet. Saved from catastrophe only by her billowing dress, which felicitously functioned like a parachute, she inspired ironic press comments and cartoons not only in Britain but abroad. The French especially seemed to relish her full-skirted skirmish with disaster. A male aviator, of course, or even those later feminine balloonists who sported bloomers as flying garb, could not have defied death in Mrs Graham's dashing manner!

Soaring and 'chuting as public entertainments remained popular until the early days of the twentieth century, and women and girl participants proved to be major attractions. Nevertheless their vivid, real-life adventures found little reflection in Victorian fiction which, in stories of hearth and home, tended to fuel female readers' domestic ambitions rather than any aeronautical aspirations they might be nurturing. Ballooning and parachuting became prominent in tales for men and boys, however, particularly when balloons began to be used over battlefields from 1876, and the British army established a special balloon section in the Corps of the Royal Engineers in 1890. Jules Verne, the pioneering science fiction writer, fictionalised aviational feats for more than thirty years and did a great deal to convince his public of the need for further research and development.

An astonishingly varied collection of male characters, from scientists and explorers to soldiers and spies, was sent aloft in short stories and novels, while comic-strips also rumbustuously celebrated the aeronautical age. Although girls might have been good for a titter in these, they resolutely concentrated on masculine ups and downs. 'The Three Beery Bounders' (by an unknown illustrator), who capered anarchically across the front page of *Funny Cuts* from 1897 to 1900, spent a lot of time in a balloon, but at the end of most episodes they walloped down to earth with a big bang or a super splash: ('"A watery grave arter all!" gurgled Flipper'). Partly inspired by their boozy antics, in 1898 the great Tom Browne sent his *Big Budget* front-page tramps, 'Airy Alf and Bouncing Billy', into balloonist action on the side of Uncle Sam in the Spanish–American War. Two years later, on the cover of *Illustrated Chips*, his other vagabond duo, 'Weary Willy and Tired Tim', ballooned belligerently against the French ('Don't you come with us, Mister Froggie, or you'll get a biff in the duff-box'). Surprisingly, Willy and Tim had discovered the aeroplane before the balloon *and* some years before the Wright brothers did so! In a November 1899 issue of *Chips*, they went up in a winged machine invented by 'the marvellous Professor Balmy' which was efficient enough to transport them at least half-way round the world.

Girls' fiction had to become airborne retrospectively. *Blackie's Girls' Annual* of the 1920s includes Winifred F. Peck's story, *The First Flight*. Set in the 1830s, it deals with mad Squire Wotton's obsession with flying. Unfortunately his methods are 'utterly empirical and unscientific', and he uses his young daughter Henrietta, 'demure as a picture in her poke bonnet, and dainty ringlets', as 'his assistant, his partner, his experimentalist all in one'. This means that the hapless Henrietta has to face ghastly hazards. It is she who goes up (and horrendously down), to break her bones in his primitive balloons and bird-machines.

There is a lot of discussion between the monomaniacal Squire and a serious young inventor named Mr Maltravers about the problems of making balloons dirigible, and of counterbalancing the weight of a human body with an instrument of propulsion, and so on – but Maltravers quickly realises that Wotton's crazy contraptions will never be airworthy. He and Henrietta fancy each other; he fears for her safety as her father comes near to completing his latest weird winged mechanism. Not unnaturally, Henrietta is reluctant to plunge in it to her probable doom, but the Squire suffers a violent stroke (triggered off by fury when he finds out that Maltravers has matrimonial designs upon his daughter), and she feels desperately guilty. Wotton, paralysed and with his speech impaired, exploits her sympathy and begs her to try out his machine while he is still in the land of the living. She agrees to make the leap, and Winifred Peck writes that 'the ghosts of all the heroes must have been near her as she ran to her appointed, unnecessary, magnificent doom'. One hopes, however, that girl readers of the 1920s would hardly have echoed this view. Henrietta's compliance with her father's cock-eyed ambition is horrible rather than heroic.

Fortunately reprieve comes in the sturdy shape of

Maltravers, who appears just as Henrietta, hopelessly flapping her plentiful feathers jumps from a tower: 'Then all the winds of Heaven and all the great waters of terror whirled in her ears as she fell', but, instead of being dashed to fragments below, Henrietta finds herself safely transfixed in mid-air. The concerned and extraordinarily competent Maltravers, perched on a nearby ladder, has managed to lasso her at the critical moment. As he carries her 'with beating heart, to his horse' she unreservedly accepts his proposal of marriage – a much better deal, at any rate, than life with her deranged daddy.

Looking back over almost eight decades, Judith Krantz's 1988 novel, *Till We Meet Again* (of which more in chapter 9), captures the excitement of a young girl's first ascent in a hot-air balloon. Fourteen-year-old Eve Coudert escapes from parental and governessing chaperonage to savour the delights of the great Air Show of 1910 at La Maladière, just outside Dijon. Aviation, of course, is still symbolic of new horizons and perspectives, both physical and psychological: from the gondola of the balloon Eve is overwhelmed by 'childlike wonder' at the endlessness of the world, and enraptured by the panorama which unfolds as she soars speedily and silently upward. Unconsciously she opens her arms 'to try to embrace the sky', and a gust of wind tears off her hat and unravels 'her inexpertly constructed chignon', leaving her waist-length hair blowing about in all directions. Embarrassed by this, and by the fact that she is recognised by a fellow passenger who is a friend of her father, and likely to report her illicit ascent, Eve nevertheless continues to thrill to the experience, feeling a link – if only briefly – 'with all the buccaneers of the sky'.

Her ecstatic fictional response is an authentic echo of that of real-life female balloonists. Iltid Nicholl, one of the founders of the Aero Club in 1901, conveyed her glowing sentiments about ballooning in an article in a magazine called *The World*:

What can appeal to a woman more than to rise for a season above the petty discontents, annoyances, and ambition of the daily round, into an atmosphere of sure beauty and serenity, which somehow or other seems usually to alter one's whole point of view, and is, therefore, wonderfully restful and refreshing? I have never been able to see why women should be incapacitated from sharing in the delights of aerial navigation. Their imagination is said to be keener and more receptive than that of men. Therefore they should be able to enjoy the fascination of the sport the more keenly.

As the article progresses, the somewhat facile feminism fades and lyricism takes over. Mrs Nicholl ends by saying:

> Another new experience is that for the first time you realise the meaning and beauty of absolute silence. On the earth, whether you are conscious of it or not, there is always, however still everything may seem, some sound. . . . But as the balloon rises all sounds seem to fall way. . . . At those great heights all is peace. There you learn to realise the 'speaking silence of a dream', the stillness of the tideless ocean that kisses the shore of eternity.

Whether or not women aeronauts had that edge on sensitivity so confidently claimed by Mrs Nicholl, it seemed to be recognised by the last decade of the nineteenth century that female parachutists drew larger audiences than men. Possibly people were more aware of the risk element when they watched someone with a pretty face and figure descend from a balloon. And there were, of course, fatalities. In 1895 at a Peterborough Bank Holiday show, Mademoiselle Adelaide Bassett's 200-foot fall to her death was witnessed by a large crowd of spectators. She and her partner, Captain Orton, planned to make simultaneous parachute jumps, but the wires became detached from Adelaide's 'chute – with tragic results, as the *Peterborough Advertiser* reported, for this 'comely and brave' lady.

This accident failed to deter other floaters and fallers, however. Only two or three days afterwards another double parachute act, Captain Spencer and Miss Alma Beaumont, made a successful jump.

London's Alexandra Palace was the launching pad for many aeronautical displays around the turn of the century. One of its great impresarios was Auguste Gaudron, a professional balloon-maker who as a young man had left France for England. A parachutist and balloonist of considerable skill, he led a small team of stunt jumpers which included both men and women. The 'Ally Pally' offered a splendid range of facilities to the public and to enterprising showmen. As well as a boating lake and a race course it provided a fairground and various arenas suitable for carnivals, military tournaments and the then extremely popular firework displays. There was also a Great Hall for musical and other performances, and a smaller theatre called the Bijou, where the celebrated Texan entertainer, Samuel Franklin Cody, drew crowds to see his Wild Western extravaganza *The Klondike Nugget*. The flowing-haired, goatee-bearded, wide-hatted Cody although apparently no relation to the original Buffalo Bill, William Frederick Cody, was best known for his emulative performances as 'Buffalo Bill' in which, sporting buckskins and silver-spurred high boots, he dazzled audiences with his flamboyant displays of cowboy riding, roping and shooting.

He was, however, also a dedicated aeronautical pioneer who, in collaboration with Gaudron and others, was able to use the Alexandra Palace's vast workshop for the development and production of kites, balloons and parachutes. It was a chance meeting with Cody and Gaudron that helped to project the seventeen-year-old Dolly Shepherd into a parachuting career. Dolly was working as a temporary waitress in the Great Hall. She had taken the job mainly in order to hear a series of concerts given by the celebrated American 'March King', John Philip Sousa; these were musical sensations for which all tickets had been sold rapidly in advance, for Sousa's dashingly brilliant marches such as 'Washington Post' and 'The Stars and Stripes Forever' had taken Britain by storm. When the first concert finished, Dolly found herself serving not only Sousa with refreshments but also the trio who joined him – John Henderson, the Ally Pally's Director of Entertainments, Gaudron and Cody.

After Sousa's concerts ended, Dolly continued waitressing, and, as Henderson, Gaudron and Cody met regularly at one of her tables, she became more and more fascinated by their conversations about balloons, parachutes and man-lifting kites. On one occasion their discussion seemed to have lost its sparkle because Cody was gloomily bemoaning the fact that the high spot of his sharp-shooting display could not take place that night. Part of his act was to shoot (blindfolded) a plaster egg from the top of his wife's head, and on the previous evening his bullet had grazed her scalp. Not surprisingly she felt unable to go on, and there was no rush of volunteers to take her place! Dolly, deeply touched by Cody's uncharacteristic dejection, found herself offering to stand in.

Cody accepted with alacrity: Dolly survived the ordeal of the performance (although its tension made one of her friends in the audience faint) and, as an expression of gratitude, Cody took her to see the kites which he was developing, and all the other aeronautical paraphernalia that was kept in the Banqueting Hall. As Dolly wrote eight decades later in her autobiographical book *When the 'Chute Went Up* (1984) this opened the door to 'adventures beyond her wildest dreams'. Gaudron took over from Cody the conducting of her tour of the workshop, and, impressed by Dolly's evident enthusiasm and the intelligence of her questions about the techniques of aviation, he asked her if she would like to make a parachute descent. Her response was emphatically affirmative. Tall, good-looking, tomboyish from childhood and with a taste for the intrepid, Dolly was an excellent candidate for a place in Gaudron's stunting team, although it was some months before he called upon her to be trained for her first drop. Not realising at the time that the vacancy had come about because one of his parachutists, Maud Brooks, had recently died from injuries suffered during a descent over Dublin, Dolly was 'bubbling with curiosity' when she met Gaudron again at the Alexandra Palace's Banqueting Hall for her first – and only – training session, which lasted no more than half an hour.

The most important part of this was learning how to fall

on landing. Gaudron demonstrated the process, and after practising several times Dolly perfected it: 'The secret was to roll on to the back as soon as the feet met the ground, and to throw the legs upwards immediately – even to go right over in a backward roll.' Obviously there was no place in this athletic but inelegant procedure for the long, full skirts of the early Edwardian era. The billowing dress which, years earlier, had saved Mrs Graham from disaster would have been inappropriate attire for the public performances of Gaudron's girl parachutists.

Gaudron explained that Dolly would wear a uniform which comprised a navy-blue knickerbocker suit adorned with gold trimmings, a rather military-looking peaked hat and high-legged boots. She thought it 'very swish'; it was certainly, for its time, daringly symbolic of liberation – both physical and psychological.

Having ascertained at their initial handshake that Dolly possessed 'the main asset of any parachutist . . . a strong grip', Gaudron took her over to a suspended parachute and instructed her to grasp and hang from the wooden trapeze bar about 20 inches long which swung below it. 'Feeling rather silly', she did so for several minutes, thus satisfying her mentor that she would be not be likely to plummet straight to her death by jumping prematurely when in the air. A 6-inch strip of webbing trailed from the trapeze bar, making a primitive sling which would rise up between the legs once the trapeze bar was clutched overhead, and eventually take the weight of the suspended parachutist's body. Dolly was undaunted by this rather basic equipment and by Gaudron's explanation that during the balloon's ascent she would have to sit balanced on the rim of the basket beneath it 'with legs dangling over the side, one hand clutching the trapeze bar and the other gripping the supporting ropes'. The basket would also carry passengers who would be more safely seated deep within it. She learned, too, that her point of landing would depend largely on the strength and direction of the prevailing wind; it would probably be about a mile from the balloon's point of departure, where the audience would wait until she was picked up by a horse and trap and returned to the enclosure to display herself triumphantly.

So far, so thrilling! The lesson ended and Dolly rushed home elatedly, convinced that her 'wildest dreams' were about to be fulfilled. She was living with an aunt in London, who owned the Ostrich Feather Emporium in Holborn where Dolly was then working. 'Aunty' had already dissuaded Dolly's parents from letting her pursue a theatrical career, and she lost no time in bringing her sky-dreaming niece abruptly down to earth: if Dolly pursued 'these mountebank stunts' she would never be allowed to enter her aunt's house again.

Aunty obviously had no time for the Edwardian 'new woman' who was demanding better educational and career opportunities, and the Vote! – and Dolly, despite her determination to join Gaudron's aeronauts, was deeply distressed by her aunt's resistance. She clung, however, to her parachuting ambitions, even though the realisation of these might result in a family rift and a great deal of financial insecurity for herself.

Fortunately Aunty mellowed somewhat during the days in which Dolly awaited her first aerial assignment. She did not exactly give her blessing to the enterprise – 'Very well – you can go up . . . but *never* talk to me about it' – but at least she would continue to provide her niece with a home and the security of regular work between her possibly spasmodic public performances. (Dolly much appreciated this, and in between jumps worked an eleven-and-a-half-hour day at the Ostrich Feather Emporium. She was certainly never lacking in energy.)

Dolly's first descent proceeded without incident. Today, when parachutists undergo fairly extensive training, it seems incredible that, after only half an hour of rudimentary instruction in the Ally Pally workshop, an inexperienced young woman on her first balloon ascent should simply be told at 2000 feet by her pilot, Gaudron, 'Get ready to jump . . . There's a nice green field over there . . . Remember how to land . . . GO!' And Dolly went, into her first fall, which was 'a heady mixture of fright and sheer exhilaration'. Even in the throes of 'a sense of elation' such as she had never known before, she remembered Gaudron's tips on landing techniques and, when the grass suddenly leapt up at her, threw herself on to her back. Her satisfaction at having broken no bones was enhanced by the unrestrained enthusiasm of the cheering throng which acclaimed her when, picked up by the pony and trap, she returned to the scene of her ascent. Additional satisfaction was provided by the – for its time – large sum of £2 10s which she received for her first parachuting performance.

Dolly quickly became known as the Parachute Queen, captivating audiences by her smart appearance both on the ground and in the air, and by her willingness to discuss with them, before and after her jumps, the thrills, chills and general sense of glamour which were still associated with ballooning several years after the Wright brothers made the first powered flight in a heavier-than-air machine in 1903.

There were, of course, several occasions when Dolly narrowly escaped death or serious injury, landing on roof-tops, in trees or a barbed-wire fence; once, her parachute failed to open until seconds before her impact with the ground, and her safe landing seemed little short of miraculous. Even worse than this 'free fall' was her experience on another ascent of 'not being able to fall at all'. In common with Gaudron's other stuntists, Dolly travelled around the country for her performances. One day, taking off from Coalville on the outskirts of Leicester, she found that things did not go according to plan. She had by now been promoted to making solo ascents as well as solo drops, going up without the presence of a possibly reassuring pilot. At 4000 feet she tugged at the ripping cord but, in spite of repeated attempts, could not get her parachute to detach itself from the balloon. There was no alternative for her but to wait until the balloon rose even higher, started to release gas through its neck and began its descent. To her horror, however, she suddenly realised that 'The flap had blown inside the neck of the balloon, virtually sealing it.' Would any gas be able to escape? she wondered. Would she *ever* come down?

Suspended for several hours in the suddenly inhospitable skies, with her clutch on life mainly dependent on the strength of her grip on the trapeze bar, Dolly rose inexorably with the balloon to a height of around 15,000 feet. At first she was comforted by the thought that she was rising ever nearer to God but, as darkness, silence, solitude, agonisingly aching arms and terrible cold began to blot out her natural optimism and, indeed, all her senses, death seemed to be very near. The only way to hold on was to keep awake and alert, and in order to do so she started to sing. She used up everything in her repertoire from hymns to popular songs ('Goodbye, Dolly, I must leave you' seemed to her to be particularly appropriate, though it was in the event ineffective!). The passage of time was measured by the changing colours of the sky, from vivid blue to sunset hues, to the blackness of night and the silver twinkling of the stars. At last, when time seemed to have lost all meaning, and consciousness was slipping into hallucination, Dolly realised that she was coming down. She landed unceremoniously at Whissendine, but the balloon immediately took off again after she had released herself, to be found three days afterwards in the North Sea with the parachute still attached to it.

Dolly rapidly recovered from this adventure, but not long afterwards faced another aerial hazard. One of her co-workers at the Ostrich Feather Emporium, Louie May, was entranced by, and longed to participate in, Dolly's parachuting exploits. Unfortunately she was to get rather more thrills than she had bargained for. Because the Mammoth balloon was damaged, she and Dolly went up suspended by separate parachutes beneath Dolly's solo balloon to make a double descent. Louie shared Dolly's fascination with the silence and exuberance of ascending, but, terrifyingly, when at 3000 feet the two girls tried to pull the cords which would release their parachutes, Louie's jammed. Neither Dolly nor she could free it, and, when they had risen to 11,000 feet, Dolly knew that no one could expect her companion, on her maiden flight, to cling to the trapeze bar for much longer. She therefore attempted, and achieved, the first mid-air rescue to be recorded. Managing to transfer Louie from one parachute to another, the determined and experienced parachutist hurtled to earth carrying the full weight of her companion, whose arms and legs were wrapped around Dolly's neck and waist.

Their landing was like 'a hammer blow' to Dolly, who threw herself backwards, while Louie bounced heavily on top of her. The girls were taken to a nearby farmhouse where Louie was found to be suffering no serious after-effects. (Nevertheless she never attempted another parachute jump.) Dolly was less fortunate, however. Her mouth was temporarily paralysed, her spine badly twisted and her legs without any feeling. At first it was feared that she would never walk again, let alone go ballooning or parachuting, but happily, after primitive electrical shock treatment and tremendous persistence with rehabilitating exercises, Dolly was able to make her aerial come-back just eight weeks after her numbing double descent.

Not surprisingly, the press made much of her pluck and presence of mind in rescuing her fellow aeronaut, although, as Dolly wryly acknowledges in her book, they glamorised the affair, presenting herself and Louie compellingly but with little attention to fact. Pictorial interpretations garbed the girls enticingly in circus trapeze artists' costumes (in which, one imagines, they might well have frozen to death while suspended for so long at a great height).

Dolly remained engagingly sanguine about all the hazards of parachuting, surviving personal accidents and hearing the news of the deaths of some of her colleagues. By 1911 she felt that the aeroplane was beginning to take over from the balloon for stunting and entertaining: 'It was as though the balloon, in many ways a symbol of the elegance and leisurely pace of Edwardian life, was being supplanted by an exciting but rather brash and noisy newcomer.' By 1912 there was also talk of war, and it seemed obvious to anyone of imagination that, if and when this happened, it would be the end of fun and innocence in aviation.

Perhaps Dolly was unconsciously motivated by thoughts on these lines. At any rate, when one spring day in 1912 she had gone up from the Ally Pally, waved her Union Jack from on high to the receding and cheering audience, and settled into the calm, swinging motion that preceded her reaching for the ripping cord of the parachute, she apparently heard a disembodied voice telling her never to come up again into the skies because, if she did so, she would be killed. She proceeded calmly with her descent, but, on getting back to Aunty's house, she rolled up her nifty aeronautical costume 'and put it in the ragbag'. Her parachuting days were over. During the First World War she became a driver–mechanic in France, and married one of her passengers, a Captain Sedgwick. Service in the Second World War included duties as a Fire Service Volunteer and an Air-raid Shelter Staff Officer.

She had the immense satisfaction, seven years before she died in 1983 at the age of ninety-six, of flying with the Airborne Regiment of the army's Red Devils. 'Seated as co-pilot, complete with headphones', she watched enraptured and envious as the team leapt into space at 7000 feet. She was particularly thrilled to meet Jackie Smith, the only female member, in whom she saw reflections of herself as a young girl 'making her determined way up the hill to the Alexandra Palace in the spring of 1903, heading for adventures and pleasures untold'.

By the end of the First World War, the heyday of para-chuting and ballooning had passed. Nevertheless stuntists of both sexes still provided entertainment and excitement at air displays, as well as setting and smashing new height and endurance records. Girl parachutists continued to have a special appeal, but although their performances were relished by audiences, these were played down in the fiction of the 1920s and 1930s. By this time, of course, despite undertaking challenging and sometimes arduous work dur-ing the war years, women had been encouraged to give up their jobs and return to their traditional places in the home. Domesticity, rather than aeronautical know-how and derring-do, was the vogue.

Paul Gallico's 'Third Time Unlucky' in the June 1935 issue of *The Story-Teller* reflects the mood. Runkleman, who orga-nises air shows and races, has problems in persuading Crack-up Kelly, one of his best flyers, to take Minorah, a girl parachutist, up for her jump. Kelly rudely expresses his mistrust of women in aviation ('The first pilot who ever taught a woman how to fly ought to be made to fly an autogiro for the rest of his life. . . . Parachute-jumpers are worse. I'd rather not have one crash from my machine!'). Neither tactful pressure from Runkleman nor the appeal of Minorah's enormous, startlingly pale blue eyes can make him agree to air-lift the girl. However, he eventually weakens after she trades insults with him and stresses her desperate need for the money she will get for making three jumps. During the flight *his* expertise and *her* ineptitude are underlined. At 3200 feet he urges her to climb out on to the wing, but once there she loses her nerve and clings on, screaming about how scared she is. All this is grist to Kelly's misogynous mill, and it is only to get away from the contempt on his face that Minorah manages at last to hurtle headlong from the plane.

Watching her descent, Kelly curses her violently for 'doing a delayed-opening drop', although this procedure naturally gives the crowd an extra thrill. He feels that the tiny, innocent-looking parachutist has fooled him, and that her fit of terror on the plane's wing was just an act. Characteristically, when he meets her later that day at a petrol station tea-shop, he tells her, 'If you ever do a delayed-opening drop like that without warning me, I'll slap you till you're sick.' He almost does, too. On the following day another pilot takes her up and, again, she keeps her parachute closed until she is only 400 feet from the ground. Kelly rushes in his car 'down the aerodrome . . . and into the billowing 'chute, spilling the air out of it'. He picks Minorah up and then, white with rage, strikes her savagely across the face, muttering, 'Damn you, Minorah, I'll kill you!'

She takes all this without a murmur (which is surprising because, like Kelly, she is Irish and might have been expected to behave as stereotypically and therefore as fierily as he.) By the time of her third drop, Kelly has learned from Minorah that she is a complete novice. With ambitions to write, she is proud, but tired, of having 'made her way alone' for three years by washing up in cafés, working in 'cheap bazaars' or 'canvassing soap'. When she heard about the local air show, she had been out of work for two months, so spent her last savings on parachutes, and contracted with Runkleman to make three jumps for the then fairly considerable fee of £30.

Torn between concern for her welfare and fury at her presumption, Kelly tries to bully her into not going up again, and resorts to 'sneaking' the truth about her lack of experience to Runkleman. Nevertheless she will not be deterred. Kelly pilots the plane for her third descent, haranguing her all the way up. When they reach jumping height he begs her to 'pull that cord' quickly once she leaves the wing, and, accusing him of being 'a thick-skulled, blind Irishman', she explains that on each of the previous drops she pulled the cord early, but the parachute failed to open until she had struggled with the emergency ring. Obviously the 'chutes bought with her pathetic last savings were a job lot.

There is a tremendous fight, with Minorah standing on the wing and Kelly, now convinced that her third drop is bound to be unlucky, trying to drag her back into the cockpit with his left hand and lashing out at her with his right. Meanwhile he has had to take the joystick between his knees – a hazardous procedure, during which the plane 'skids and lurches'. Minorah twice bites his wrist but still he hangs on, threatening to hit her next time with a spanner. She still wants to jump, to earn that much-needed £30, so Kelly plays his last card to prevent her: he yells, 'God help me, Minorah, I love you . . .'

By this time the plane has descended too low for a jump anyway; their landing is rackety but safe; they fall into each other's arms with tears and kisses, and Kelly proudly tells Runkleman and an approaching ambulance driver that Minorah is to become Mrs Kelly and that she has made her last jump. One suspects, somehow, that she has also had her last stab at any kind of career – Kelly will see to that! She is certainly unlikely to be allowed by him ever to trespass again on the 'male' preserve of aviation.

Generally speaking, during the period between the two world wars, female career aspirations found more vigorous expression in juvenile than in adult fiction. The twopenny weekly *Girls' Crystal* produced a particularly satisfying bunch of spirited, competent and strong-minded heroines. The colourful exploits of Pat Lovell, a teenage girl reporter, ran as a series for several months, and *Pat's Daring Parachute Jump* (published in May 1939 during the run-up to the Second World War) promised the reader plenty of thrills. Appropriately, it includes some foreign 'baddies' who try to keep Pat at a distance from their activities; they are, of course, spies. Their mission is to steal the prototype of a 'self-opening parachute' that is being developed in Britain by an inventor called Frazer. His daughter Betsy is his test pilot or jumper, and Pat first encounters her when she is hanging from a tree after a drop. Despite her own advanced ideas about careers for girls, Pat is surprised to find that the parachutist is female 'although doing a man's work'.

As a friendship builds up between the two girls, Pat realises that Betsy is losing her nerve. She is due to make the final test of the automatically opening 'chute which would 'make the toll of life in aeroplane accidents almost negligible . . . and was so compact that every aeroplane passenger could have one'. To bolster up Betsy's spirits, Pat agrees to accompany her on the flight for the final test – though not to make the jump, the thought of which thoroughly scares the journalist, even though she has undertaken many intrepid assignments during the series in her efforts to find good stories for her paper.

Fortunately Pat wears one of the marvellous new para-chutes too – just in case the foreign spies might have tam-pered with the aeroplane. She ends up having to make the jump (from 10,000 feet) instead of Betsy, who has been abducted by the villains. Aware of the crowds of sightseers and pressmen below, and what successful publicity for the parachute will mean to the Frazers, Pat overcomes her fear sufficiently to throw herself out of the machine: 'She was diving down through . . . the clouds, with the wind whistling and screaming in her ears . . . to patches of green below, to dots, specks, that grew larger and larger and larger, and nearer.' Happily, 'the parachute, without her doing anything to help it', opens in time for her to make a safe landing.

Betsy is rescued from the baddies, the Frazers' fortunes are assured, Britain will have the use of their wonderful new parachute in any future war, and Pat has a sizzling story for her newspaper. Nevertheless, when 'with eyes shining and cheeks flushed' she receives the congratulations of her colleagues, she confesses that she 'wouldn't jump again for a million pounds'.

As we shall see, heroines juggling with the joysticks of aeroplanes were to become far more popular fictional subjects than girl balloonists or parachutists. Aeroplanes gave writers more scope – despite the fact that Dolly Shepherd and some of the other early parachutists had adventures which were almost more vivid than fiction. Reporter Pat's exploits were chronicled by 'Elizabeth Chester' (in reality E. L. Rosman). He was one of the Amalgamated Press's team of male authors who created millions of words, thousands of lively heroines and the generally progressive mood of girls' weeklies during the 1920s and 1930s. Editorial policy was that men would produce more exciting fiction for girls than women writers were likely to do; it was felt that the adult female's maternalism and protectiveness for young girls would restrict the circumstances which they might create for their heroines, and inhibit the action of plots. As Ida Melbourne, Rosman also created a pair of 'flying sisters' for the *Girls' Crystal*'s companion paper, *The Schoolgirl* (see chapter 8).

In real life, parachuting as sport and as entertainment has undergone a resurgence in the 1980s and 1990s. Those who make jumps for their own satisfaction, and under sponsorship for charities, include many women. For some time, too, there have been female performers in those groups who have carried free-fall and formation skills to their zenith – the army's Red Devils, the Metropolitan Police Parachute Team, and the RAF's Falcons.

After meeting some of these 'young adventurers', nonagenarian Dolly Shepherd, the early Parachute Queen, commented:

Although the equipment they use bears little resemblance to my old 'limp parachute' suspended beneath a balloon with its trapeze bar and sling, and although the skydiving skills now enjoyed are beyond anything that I dreamt of in my Edwardian days, the sensations and the pleasures and the rewards are still the same. The sheer exhilaration, the freedom of the skies, the joy of drifting earthwards under a smiling canopy – these have not changed.

Chapter Two

From Ballooning to Barnstorming

The aeroplane soared high into the clear air.   Under Peggy's skillful
hands the plane fairly flew.

(An illustration from *The Girl Aviators on Golden Wings'*, 1910).

In aviation there seems no place for the woman engineer, mechanic or flier. Too often, people paid money to see me risk my neck more as a freak – a woman freak pilot – than as a skilled flier.

BLANCHE STUART SCOTT (1916)

RATHER more staid than Dolly Shepherd, Gertrude Bacon was an early British balloonist, but not a parachute-jumper. Recruited for sky trips by her father, the Reverend John M. Bacon, she shared his serious approach to aeronautics and his desire to use ballooning as a means of furthering scientific knowledge.

Bacon's enthusiasm for ballooning began in 1888 when he made his first ascent. Soon after this he abandoned his part-time ecclesiastical career because he felt that the Church was trapped in a rigid and bigoted conventionalism which frowned upon scientific exploration. He claimed relationship with Roger Bacon, the scholarly Franciscan monk, who as long ago as the thirteenth century had projected a potent image of a method by which men might fly: 'It is possible to make engines for flying, a man sitting in the midst thereof, by turning only about an instrument, which moves artificial wings to beat the air, much after the fashion of a bird's flight.' Not quite an accurate conception of the aeroplane – but as he also predicted that a hollow metal vessel filled 'with aetherial air or liquid fire' would one day float through the skies, he came pretty near to envisaging hot-air and gas balloons.

John and Gertrude felt that they were following in Roger's progressive footsteps, as well as in those of the great sixteenth-century philosopher Francis Bacon, who, they liked to think, was another of their 'illustrious kinsfolk', although Gertrude admitted that such a connection, 'well-attested' as it was, could be only remote. She was to describe her father's achievements in 1907 in a loving (and extremely long) book entitled *The Record of an Aeronaut*, and seems always to have been deeply influenced by him.

John Bacon had taught his children at home, concentrating on the study of languages and science. His particular passion for astronomy was one which Gertrude shared, becoming in 1890 at the age of sixteen the youngest 'Original Member' of the British Astronomical Association. She also participated in many of her father's other activities; firework-making was a favourite pastime, and the list of John's and Gertrude's interests included printing, cycling, photography and handbell-ringing. They arranged regular flower shows as well, to encourage the 'unsophisticated rustic dwellers' in their Berkshire village to transfer their energies from 'the public house and the illicit snaring of game' to flower, fruit and vegetable growing.

Gertrude first went up in a balloon in 1898, ten years after her father's first ascent. She was twenty-four, and had for several years been his unofficial assistant and hostess to his gatherings of astronomically and aeronautically inclined colleagues. She made three sea trips with him, to Norway, India and America, for the purpose of studying and photographing eclipses.

Father and daughter ascended from the Crystal Palace and travelled over Hertfordshire and Cambridgeshire. Gertrude's 'first experience of cloudland' was an amalgam of satisfaction and frustration. Once aloft, a thick haze made photography impossible. Gertrude had been concerned that her poor head for heights might have ruined the trip for her but, in the event, she found no problem in gazing down from the balloon at the changing panorama of buildings, fields, woods and rivers; it seemed to her that these and not the balloon were moving.

Gertrude's first landing was an infelicitous affair, which emphasised that, despite a great deal of improvement in design and manufacture over the years, the balloon was still extremely vulnerable to the vagaries of wind and weather. The newly reaped cornfield chosen for the Bacons' descent proved to be less accommodating than it seemed from 'a mile's elevation'. 'A stunning crash announced their arrival on terra firma'; the prongs of the grapnel which should have embedded in the ground to anchor the balloon failed to do so because prolonged drought had baked the earth iron-hard, and a treacherous wind bounced the balloon and its crew over the fields – scattering corn sheaves and farm animals in their wake – towards a deep cutting of the Great Northern Railway. Just before reaching this, the balloon was blown into a double row of telegraph and telephone wires which finally grounded it.

Some time afterwards, in November 1899, Gertrude made an especially memorable flight. She accompanied her father and Stanley Spencer, who shared with his brother Percival a long-standing interest in, and knowledge of, ballooning in all its aspects. They went up prosaically enough from the grounds of Newbury gasworks through drizzle and a cloudy night sky, but rising above this they found themselves in what Gertrude described as 'the unutterable beauty of fairy-land'. The main purpose of the trip was to observe 'that great shower of meteors' which were predicted to radiate that night from the constellation Leo. The meteors appeared only three times in each century, and, with half England resolving to sit up until the small hours to see them, Gertrude felt privileged to have the opportunity of watching the heavens from the vantage-point of a balloon.

Bacon was to report the spectacle for *The Times*, and Gertrude was entrusted with making a photographic record. (Her pictures, taken with a plate-camera on many ballooning occasions, were impressive; they ranged from studies of cloud formations to unusual aerial views of famous London sights, country backwaters and cricket matches.) Although the shower of meteors failed to put in an appearance, in Gertrude's view more than adequate compensation for their absence came in the breath-taking vistas of the night sky. As she wrote afterwards in *The Record of an Aeronaut*:

For us alone Sirius flashed magnesium blue, and the other stars glistened as jewels in a blue-black velvet sky. For us, and us alone [the] filmy, tossing billows of clouds turned to silver in the moonlight, whose deep hollows harboured shadows of richest purple, whose boundless, snowy expanse stretched to the horizon's limit in one vast, silent, glorious ocean.

She was soon to discover that 'fairyland had proved a trap, and meteor hunting had brought us woe'. The balloon behaved waywardly, continuing to rise when it should have started to descend. In what turned out to be a ten-hour flight, Gertrude and her companions realised that they were being swept inexorably westwards, and feared that the balloon might eventually come down some distance across the Atlantic. (They had packed a great deal of equipment into their 6-feet long, 3½-feet wide and 3-feet deep basket, but life-jackets and parachutes were lacking.) At last their craft began to come down, at first slowly and then suddenly, caught by 'a wild gust' of wind which smashed the balloon to earth, dragged its occupants through a barbed-wire fence and then enmeshed them at the top 'of a weather-beaten oak tree'. John Bacon's leg was badly torn on the barbed wire, Gertrude broke her arm and fainted.

None of this put them off future flights, however. The pinnacle of Bacon's career may well have been his reading a paper on 'The Balloon as an Instrument of Scientific Research' to the Royal Society of Arts. It would be difficult to say which of Gertrude's exploits was the most outstanding, because so many aviational opportunities came her way. Bridging the gap from ballooning to powered flight, at Stanley Spencer's invitation in 1904, Gertrude became 'the First Woman in the World to Make a Right-away Voyage in an Airship'. She experienced all the excitements and expansiveness of the air shows which began to take place in Britain and France after the Wright brothers had made their pioneering flight in 1903. At the first international aviation gathering at Reims in August 1909, Gertrude became the first Englishwoman to fly in an aeroplane when Roger Sommer took her up in a Farman biplane.

She never became a flyer in her own right, although for some years she was to produce articles and to lecture on aviation. Ultimately she was able to add to her achievements those of being the first British woman to travel on a seaplane and on a commercial flight from London to Paris. One feels that John Bacon, who died in 1904 just after the era of the aeroplane began, would have been proud of her. Possibly held back by excessive admiration for her father and dedication to continuing his work, Gertrude did not marry until she was fifty-five. She died at the age of seventy-five in 1949, having never lost her interest in the ever-proliferating forms of flight that had developed since her early ballooning days.

Even at the time when the Bacons were making their early ascents, determined experiments with powered and winged flying-machines were being made in several parts of the world, notably in France and the USA, where towards the end of the nineteenth century the race for leadership in the field of aviation was definitely on. From the middle of the nineteenth century dirigible airships had been developed from basic ballooning. These were successively steam-, electric- and petrol-driven, but although many pioneers felt that the future of aviation was in the airship, there were still problems of controllability and limitations in their lighter-than-air structure.

Otto Lilienthal, the outstanding German gliding pioneer, was experimenting with a carbonic acid gas engine just before he died as the result of a glider crash in 1896. In Britain three years later Percy Pilcher's progress with the development of oil-powered flight was also abruptly curtailed when he died after a gliding accident. There is no doubt that several aviators on both sides of the Atlantic were on the verge of achieving controllable, powered flight when Orville and Wilbur Wright successfully tested their *Flyer* on 17 December 1903, at Kill Devil Hills, Kitty Hawk, North Carolina, achieving a twelve-second flight of 36.5 metres. They made three more flights on the same day, the longest of which covered 260 metres and lasted for fifty-nine seconds. Sustained, controllable, man-carrying, powered flight, which had been sought by so many for so long, was thus demonstrated.

Over the next decade – which turned out to be the run-up to the First World War – technical developments, stunts and showmanship, and the making and breaking of records, proceeded with bewildering pace and variety. Women as well as men played their part in these feats of aviation. In 1908 Madame Thérèse Peltier became the world's first female aeroplane passenger. She travelled with Léon Delagrange in his Voisin on a 150-metre flight over Turin. She also achieved the distinction of becoming the first woman ever to make a solo flight, although she was never a qualified pilot. In England the first woman passenger was actually an American, Isabel Cody (the wife of the Wild West showman and aeronautical pioneer), who was taken aloft by her husband in July 1909 over Laffan's Plain, Hampshire, in the *British Army Aeroplane No. 1*. Surprisingly, perhaps, women in America had to wait a few weeks longer before going up in aeroplanes. The first of them, Mrs Ralph van Deman, became one of Wilbur Wright's passengers in October 1909.

Only a few months later, women began to qualify as pilots. In March 1910, the twenty-three-year-old Raymonde de Laroche was the first to be licensed. There was no false modesty about this strong-minded Frenchwoman who claimed to be a baroness as well as a gifted artist and actress, and who became the first feminist spokeswoman in aviation. In support of her conviction that women could pilot planes just as well as men, she declared that flying 'does not rely so much on strength as on physical and mental coordination'. Sadly, in 1919, like so many early flyers, she was to be killed in an aircraft accident after only a brief career.

The first British woman to make a solo plane flight was similarly ill-fated. Miss Edith Maud Cook was already an experienced professional parachutist, jumping under the name of 'Viola Spencer'. Although unlicensed, in 1910 she began to pilot Blériot monoplanes with the Grahame White Flying School in the Pyrenees, using the name of 'Miss Spencer Kavanagh' for these aerial acts. Later that year she lost her life after a jump from a balloon over Coventry.

The first American woman known to have flown solo was Blanche Stuart Scott who, like her French forerunner, Raymonde de Laroche, felt that female flyers could be as adept and intrepid as men. She was accepted for pilot training by Glenn Curtiss (founder of the first US aeroplane company in 1907 and the first American, after the Wright brothers, to fly). However, despite his progressive attitude towards aviation, he firmly felt that women's place was on the ground. His agreement to train Blanche was somewhat reluctant: he put a throttle-block on her plane to prevent it from taking off, but, in September 1910, despite this block and Curtiss's instruction that she was not to leave the ground, Blanche managed to climb to 40 feet. It is to Curtiss's credit that he allowed her to continue her training and to become part of his exhibition team, in which, at the beginning of October, she made her first public flight at Chicago. Dubbed 'the Tomboy of the Air', she was to become a brilliant stuntist, sometimes earning $5000 a week, but the serious career in aviation to which she aspired was denied her. After six years of providing thrills for audiences throughout America by skimming under bridges, making death-defying dives and flying upside down only 20 feet above the ground, Blanche abandoned aeronautics at the age of twenty-seven with a comment (quoted at the beginning of this chapter) which many subsequent female flyers might have echoed with fervour.

Despite Blanche's first flight, the official credit of becoming the USA's first woman pilot went not to her but to Bessica Raiche, *née* Medler, who was a person of many talents and much resource. She drove a car, and excelled in sports, music and modern languages. She went to Paris to further her musical studies, and became intrigued by the feats of 'Mme la baronne de Laroche', returning to America with not only an interest in flying but a French husband as well. Together, in their Mineola, New York, living-room, they constructed a small and extremely fragile plane in which Bessica was to make her original solo flight on 16 September 1910 – although she had never been up before, nor indeed had received any aeronautical tuition. Improving her performance during the following weeks, she received from the Aeronautical Society a diamond-studded gold medal inscribed to 'The First Woman Aviator of America'.

Bessica was not only an aviational but a sartorial pioneer. Quickly discovering that long skirts could create hazards in flimsy planes, she adopted riding-breeches as flying garb, a fashion which several celebrated aviatrices, including the great Amelia Earhart, were to follow. Bloomers, of course, had been around, if not exactly in vogue, since the 1850s, and Bessica had already taken to these for sporting and other ground-based activities. She and her husband expanded their home-based silk, wire and bamboo aircraft industry into a profitable French–American company. Eventually, when Bessica had to give up flying for health reasons, she embarked upon another demanding career, becoming a doctor of medicine.

American publishers lost no time in reflecting in fiction the feats of their early women flyers. In 1910, the year when Blanche Stuart Scott made her first unorthodox flight, Messrs M. A. Donahue & Co. of Chicago and New York published the first of their Girl Aviators series by Margaret Burnham. This series was heralded by Donahue as 'just the type of books that delight and fascinate the wide-awake Girls of the present day who are between the ages of eight and fourteen years'.

The same publisher had already provided other series books. Those specifically for boys included The Aeroplane Series, Motor Boat Boys, Radio Boys, Victory Boy Scouts and the American Boys' Sports Series. For girls, motoring (the Motor Maid Series) had preceded aviation as an exciting fictional subject: further established series were Girls' Liberty (misnamed, and consisting mainly of reprints of Mrs Molesworth, Mrs Ewing, L. T. Meade and other fairly staid Victorian children's writers), and, far more in keeping with early twentieth-century images of 'the New Woman', the Campfire Girls Series. There was also the beginning of an American version of a school series 'dealing in an interesting and fascinating manner with the life and adventures of Girlhood' in the shape of the adventures of a girl called

Peggy Parson. (This obviously did not catch on with American readers despite the tremendous popularity in Britain, from 1906, of Angela Brazil's tales of spiffing and sporty schoolgirls.)

Margaret Burnham's Girl Aviators Series had run to four titles by 1911. The first story, *The Girl Aviators and the Phantom Airship*, introduces the two – apparently teenage – girl flyers, Peggy Prescott and Jess Bancroft. Each has a pilot brother – Roy and Jimsy, respectively. Peggy's and Roy's father has invented a 'non-capsizable aeroplane of great power', the plans of which he confides to them on his deathbed. The rest of the book is mainly concerned with the efforts of a couple of baddies to force Mr Prescott's offspring into handing over the plans. Peggy and Roy manage to hang on to – and develop – these, despite kidnapping and the weirdly murky manoeuvres of a 'phantom' aeroplane which is supposed to terrorise them into submission. Peggy is a backer-up of her brother rather than a prime mover in this adventure, but in the second book of the series, *The Girl Aviators on Golden Wings*, her flying prowess is recognised by the young protagonists as being even greater than that of the boys, Roy and Jimsy. She 'saves the day' more than once, despite occasional anti-feminist offerings – 'This is not work for women or girls!' – from adult friends and relatives. (Incidentally the book is aptly titled; aeroplanes of this 'string and stick' vintage, with their doped canvas bodies, really do glow golden in the air if the light is good.)

Jess, though slightly less capable than Peggy, is also able to pilot a plane. (No mention is made of training, licensing or possible age qualifications, but perhaps this is in keeping with what Ruth Law, a contemporaneous real-life aviatrix, de-scribed as 'the good old crazy days of flying'.) The girls' gear, whenever they clamber into one or other of their mono-planes, is apparently far from fetching: they 'slip into linen coats and don their hideous masks and blue sun-goggles' with never a qualm, as prettiness properly takes second place to professionalism.

The Girl Aviators on Golden Wings is set in 'The Great Alkali' on the edge of the Nevada desert and, at first, it is a Western rather than a flying story, focused on a gold-mine recently discovered by a Mr Jim Bell. Conventional foul-mouthed, cross-eyed or greasy Mexican bad guys trail Bell and the trustworthy teenage chums on horseback across scorching wastes, hoping to discover the mine's location and to file a claim for it as their own.

Jim Bell has 'the happy idea of conveying the precious product of his mine by aeroplane'. Peggy, Roy, Jess and Jimsy are to ferry the gold across the desert, which 'automobiles cannot cross', while 'transportation by wagons would have been prohibitive in cost, as well as almost impossible to achieve'. Nevertheless, Mr Bell and the young aviators have first to cross the Alkali by horse and wagon in order to convey themselves, their personal effects and three boxed and unassembled planes to the mine. Bell has tremendous faith in the capabilities of his young employees; this appears to be justified, because on arrival at their destination they manage, with astounding speed, to put together their fragile craft and to make them absolutely airworthy. The narrative provides frequent reminders of the flimsiness of aeroplanes at this time, with Peggy demonstrating her mechanical skills when their simple wood and canvas frames or 'intricate' innards need emergency repairs. (Her expertise in this field is described in a fairly low-key way, but early aviatrices found that the public was even more impressed and surprised by their engineering abilities than by their flying and stunting feats.) On one occasion after a crash-landing, Roy, who has been piloting, is 'pain-crazed' from a dislocated shoulder. Incapable of attending to the repair of the damaged plane, and under great stress, he abandons his generally progressive attitudes and, when Peggy says she will put things right, grumbles that this '*is* a man's work'. However, he nobly eats his anti-feminist words when 'Sis', after identifying the problem (a radiator leak) makes nifty and effective repairs by filling the hole with some well-worked chewing-gum.

A little later on, in a wild race against the villains to file the claim to the gold-mine, plucky Peg has to fly a monoplane at its top and terrifying speed of 60 miles per hour. After an enforced landing she is temporarily unable to take off because, at a time when any aeroplane was an object of tremendous curiosity,

> the crowd stupidly clustered about it like bees round a rose bush. The delay was maddening, but Peggy dared not start for fear of injuring someone.
>
> 'Won't you please stand aside?' she begged for the twentieth time, but the crowd just as obstinately lingered.
>
> Suddenly an idea came to her. She cut out the mufflers and instantly a deafening series of reports, like a battery of Gatling guns going into action, filled the air. . . . The inhabitants of Blue Creek literally tumbled all over each other in their haste to get out of the way. Five seconds after the deafening uproar commenced a clear path was presented, and, before the crowd could get used to the sound and come surging around again, Peggy started the aeroplane up. Amid a mighty shout it took the air and vanished like a flash in the gathering dusk. The race against time was on.

Peggy's use of aeronautical know-how to deflect those who impeded her plans presaged similar and even more dramatic exploits by later fictional aviatrices. Captain W. E. Johns's Worrals in 1941 (see chapter 9) and the Mills & Boon heroine Leigh Bishop in 1982 (see chapter 12) had to press their technical skills into action to foil evilly intentioned passengers. Both girls were in direr circumstances than those with which Peggy deals so crisply: Worrals was threatened with torture by the Gestapo, and Leigh with rape at 6000 feet.

The tone of Margaret Burnham's pioneering Girl Aviator books is often surprisingly sedate. Peggy's and Jess's chilling skirmishes with their opponents, or their mad dashes through the skies, are clocked up simply as 'events fraught with great importance to our young adventurers', etc. However, their flying of fragile aircraft over deserts which defied other forms of transportation, and their manipulation of them through wayward air currents, gunfire from rivals' planes and other hazards provided plenty of thrills. Most important, perhaps, was that the stories were accepted as echoes of the achievements of real-life women pilots, underlining the fact that daring young girls in their flying-machines were not merely a nine days' wonder.

In Britain female flyers were still not regarded as suitable subjects for fictional treatment, although early in 1911 the romantic weekly *Sunday Stories* featured an airborne elopement. However, even when 'The Flight of Kitty Smart' allowed its eponymous heroine to take to the skies, the mood was somewhat archaic, with Kitty declaiming to her 'Mr Right', 'Save me! I can hear footsteps. For love's sake, take me out of this on your aeroplane!' It is, of course, *he* who pilots the biplane, with Kitty as his willing but wilting passenger.

The earliest Girl Aviator stories were published before August 1911 when Harriet Quimby became the first American woman to receive her pilot's certificate. She was followed soon afterwards by Matilde Moisant. Both had been trained at the Hempstead, New York flying school established by Matilde's brother John, whose death in October 1910 during a flying demonstration had saddened but not deterred the two budding aviatrices. They became members of the Moisant exhibition team which in November 1911 went to Mexico, where Harriet and Matilde achieved notability as the first women ever to be seen in the skies. At an air show there to celebrate the inauguration of President Madero, Harriet almost came to grief. Her engine failed just 150 feet above ground, but fortunately she managed to clear various obstacles and to avoid landing until she came to a safe, smooth area.

Ever eager for publicity, Harriet published an account of aspects of the Mexican trip in the New York publication, *Leslie's Weekly*. Already the paper's dramatic critic, she was an established journalist before she took up flying as a second career. By 1902 she had become a staff writer on the San Francisco *Dramatic Review*, as well as being a regular contributor to two other of the city's journals. Glamorous, and well known in San Francisco, Harriet was considered one of the local 'beauties', and her portrait hung in the elegant and prestigious Bohemian Club on Nob Hill until the premises were destroyed in the 1906 earthquake.

By 1910 she was thirty-five – ten years older than she admitted to. She had always hoped to retire at this age with plenty of money behind her, and, although journalism had not provided this, she hoped that aviation would quickly bring sufficient funds to ensure freedom from routine work. Only two days after receiving her licence she had earned $1500 for a headline-making moonlight flight over New York's Staten Island. She was soon earning $500–600 a time at air races and exhibitions, but ambition urged her towards a bigger, brighter and more international prize. Deciding that Europe rather than America was the focus of aircraft manufacture and air racing, she made up her mind to become the first woman to fly the English Channel, and persuaded *Leslie's Weekly* to sponsor her attempt. Blériot's famous first cross-Channel flight had taken place only two years earlier, and the few flyers who had since succeeded in emulating him had all been male. Once Harriet arrived in England in March 1912, she discussed her plans with the editor of the *Daily Mirror*, who agreed that his paper would co-sponsor her.

Handsome, dark-haired and striking, Harriet was as colourful in her personality as in her appearance. Claiming a wealthy background and a private education on both sides of the Atlantic, in reality she was almost certainly the daughter of a failed Michigan farmer. However, elements of mystery and contradiction about her early life added spice to her public image. Her flying garb was vivid and unusual; she had designed it to conceal the fact that she was a woman because, apparently, she had not wanted her newspaper bosses to know about her flying lessons until she had achieved her licence. Her dress of plum-coloured satin was long and enveloping, but could rapidly be converted to pantaloons, and the outfit was topped by a face-and-hair-hiding hood. Even when she felt that there was no further need for concealment, Harriet continued to wear this, or similar plum or purple attire, whenever she flew.

She naturally looked forward to enhancement of her

image, as well as to financial reward, once she had made her Dover to Calais flight in a monoplane provided especially for the purpose by Blériot himself. By a quirk of fate, however, reportage of her achievement was to be considerably less comprehensive than she had expected. The news of the sinking of the *Titanic* broke on 17 April and, of course, as well as monopolising almost every newspaper headline and front page in England and America, it took up a great deal of inside space. Harriet's departure from the celebrated Kentish white cliffs in the early hours of 16 April and her arrival at Cape Gris-Nez some 25 miles south of her expected destination were relegated to short, almost hidden-away accounts. However, the sponsoring *Daily Mirror did* provide more adequate coverage, while *Leslie's Weekly* published Harriet's own report of her historic feat.

Almost until the moment of take-off she had lived in trepidation of being forestalled. Eleanor Trehawke Davies, a wealthy Englishwoman with a mania for flying (fortunately, for Harriet, as a passenger rather than as a pilot), crossed the Channel only days before Harriet's attempt, travelling – just as Harriet planned to do – in a Blériot monoplane, flown by her protégé, Gustav Hamel. This young man, the son of a British-naturalised German doctor, was recruited to Harriet's team. Though generous with advice, he was convinced that no woman could success-fully make the Channel flight. He devised a scheme whereby he would don Harriet's concealing, hooded, flying outfit to make the crossing. She would await him at a pre-arranged spot in France, switch clothes, fly the Blériot on to Calais and claim the honours. Of course the stalwart Harriet would have none of this; she did, however, allow him to tie a hot-water bottle around her waist as she settled into her open cockpit, and was no doubt grateful for this, because, although she was wearing two layers of silk underwear beneath a wool-backed purple satin suit, and a coat and mackintosh over it, fog and 'bitter cold – the kind of cold that chills to the bones' very soon engulfed her.

As well as cold and poor visibility, Harriet had to cope with a near mishap which she described in her *Leslie's Weekly* account of the flight: 'The machine tilted to a steep angle, causing the gasoline to misfire. I figured on pancaking down so as to strike the water with the plane in a floating position. But, greatly to my relief, the gasoline quickly burned out and my engine resumed an even purr.'

Although her flight did not produce the immediate rush of acclaim which she had expected, Harriet's achievement gained fuller recognition soon afterwards, particularly in America. Some aviation correspondents in England, however, remained firmly unimpressed. The much-respected journal *The Aeroplane* (whose editor C. G. Grey still denigrated female flyers even a quarter of a century later – see chapter 9) grudgingly allocated twenty lines to a description of her feat. Although describing Harriet as 'a woman of unusual initiative, determination, and ability', the article was damning about her technical prowess: 'her daring, or recklessness, may be gauged by the fact that she used habitually to fly up to 2000 feet without knowing how to glide, and it never seemed to occur to her that she might have difficulty if her engine stopped.' An English aviator who saw her first attempt at a glide was quoted as saying that 'She started it at 1000 feet and practically "pancaked" the whole way down in spasms.'

Such chauvinistic reporting could hardly have endeared the British to Harriet, and it seems that she had a further reason for resenting her treatment at their hands. Gustav Hamel's help was acknowledged in her several accounts of her cross-Channel flight, but in her *World Magazine* article there is an indication that she saw him as deliberately undermining her efforts. According to this, he flew Eleanor Trehawke Davies across the Channel *after* learning of Harriet's plans and agreeing to become a consultant in her team. This, coupled with his suggestion that he should surreptitiously pilot Harriet's Blériot on the great day, has led to speculation that Hamel viewed with disfavour the idea of an American rather than a British woman becoming the first to fly over those famous 21 miles of water.

Miss Trehawke Davies appears to have been unruffled by either Harriet's rancour or Hamel's wiles. She not only made

47

several subsequent Channel flights with Hamel but, also as his passenger, became the first female in the world to experience looping in an aeroplane. This took place over Hendon in 1913, precursing by some two years the first solo 'loop the loop' performances by the American aviatrices Ruth Law and Katherine Stinson. In fact Eleanor Trehawke Davies seems to have had a charmed aviational life, surviving many crashes, and afterwards always being able to afford new, replacement planes – and pilots.

After her return to the USA, Harriet Quimby was much in demand at air shows and exhibitions. She was meticulous in checking mechanical details before taking off on any of her flights, and claimed that this was why she never had a crash. On 1 July 1912, only a few weeks after her historic flight, she planned to fly her new white Blériot round the Boston Lighthouse at a Squantum Airfield show. She took up with her the event's organiser, William Willard, a hefty individual, who was earnestly admonished to keep still in the plane because sudden movement of his weight might upset its balance. At first all went well, and the crowds packed in the stands overlooking Dorchester Bay saw Harriet fly out and circle the lighthouse. To their amazement and consternation, however, on its return the plane suddenly went into a dive, and first Willard, then Harriet, was thrown out. They landed in the shallow waters of Boston Harbour and died instantly, with their backs broken. Strangely, despite the concern with safety which made Harriet check her aircraft so thoroughly before each flight, on this occasion neither she nor Willard was strapped in. With hindsight it seems incredible that they did not wear seat-belts, but these were unpopular with many early flyers. The bizarre nature of Harriet's death was underlined by the fact that her Blériot righted itself after the dive and landed in the water only 100 metres from where Harriet had fallen. Although floating upside down, it had suffered only slight damage.

Harriet had a flair for projecting a glittering image and attracting publicity, but there is no doubt that she took flying seriously. Her tragic early death was a great loss to aviation and a terrible shock for her friend Matilde Moisant, especially as it followed fairly close upon her brother's death during an aerial performance. Matilde's own achievements included the winning of the Rodman–Wassamaker altitude trophy in September 1911. In spite of several accidents, she remained a fearless pilot, and agreed to give up flying at the age of twenty-six only as the result of parental pressure following John's and Harriet's fatalities. She agreed that a forthcoming flight in Texas would be her last – and, in the event, it almost ended her life as well as her aeronautical career. In common with the later celebrated American pilot, Jacqueline Cochran, Matilde believed that thirteen was a lucky number. However, when she landed her Blériot, *Lucky Thirteen*, it caught fire as its fuel tank sprang a leak. Almost miraculously she was able to escape unhurt, possibly saved from being badly burned by the heavy coverage of the tweed knickerbocker suit which she favoured when flying. Years later – shortly before her death in 1964 – Matilde commented: 'My flying career didn't last awfully long because in those days that was man's work, and they didn't think a nice girl should be in it.'

One of Harriet's and Matilde's contemporaries, Ruth Law, had a rather longer career in aviation, with the encouragement of her 'trick-parachutist' brother, Rodman Law, and her husband and business manager, Charles Oliver. Although she appreciated the value of stunt flying in building up a career – and an income – she aspired to push out the frontiers of aviation by bettering the established achievements of male as well as female pilots. In 1916 she broke the American cross-country record in a non-stop flight from Chicago to New York. To make this possible she had ingeniously mounted her maps in strips on to a roller which she tied to one of her knees, thus being able to wind it on without having to let go of the controls. Previously she had set a new women's altitude record of 11,200 feet. She had also given joy-rides and breath-taking displays of night flying.

Despite her widespread fame and popularity, Ruth was unable to obtain a flying job with the American services during the First World War. (She was not alone in this; see

chapter 3.) Contenting herself 'for the duration' with giving
exhibition flights as fund-raisers for the Red Cross, after the
war she and her husband ran an elaborately structured flying circus. Her stunts were spectacularly daring and skilled,
but when she stood in the centre of a biplane's wing while its
pilot looped the loop three times, her husband, fearful for
her life, precipitately announced her retirement to the
press. Ruth concurred with some regret, bemoaning the fact
that flying had become hemmed in by 'so many rules and
regulations'.

Although Ruth Law was hailed as the first woman pilot to
loop the loop, her compatriot, Katherine Stinson, was to be
similarly acclaimed. She carried this skill to extraordinary
lengths, in 1915 apparently making eighty consecutive loops
before flying upside down for a brief period and then carrying out several spins. Katherine was a member of a remarkably air-minded family. She originally trained to fly at the
age of sixteen, in hopes of earning a large income which
would finance her musical studies; however, flying soon
became an end in itself. In 1913, shortly after completing her
training, she and her mother, Emma, founded the Stinson
Aviation Company. The enterprising pair managed to do
this with quite limited capital, borrowed from friends. As
well as helping to organise the manufacture and sale of
planes, Katherine began to give exhibition flights, and in
1913 became the first woman to carry mail by air.

Fully aware of the hazards of her chosen career Katherine
tried – unsuccessfully – to dissuade her sister Marjorie from
flying. Both Stinson sisters were to become celebrated. They
taught their brother Jack to fly, but neither girl felt that their
other brother Eddie, who was undisciplined and a heavy
drinker, should take to the air. (Nevertheless he did so,
arranging his own flying lessons and eventually teaming
with Jack to form the Stinson Aeroplane Co. in Dayton,
Ohio.) Katherine and Marjorie next established a flying
school. Katherine's exhibition flights brought in the money
initially required for this, and Marjorie became its principal
instructor.

Katherine's remarkable achievements included a tour in

1917 of China and Japan, countries in which no women had previously flown and where the response of enormous audiences was ecstatic. Her stunting at night, when she carved letters in the sky from fireworks and traced the pattern of her dramatic dives and spins with magnesium flares, earned her the title of 'Air Queen'. Apparently she also became a symbol of emancipation for Japanese women and girls.

Katherine's dazzling feats were eventually ended in 1920, when tuberculosis and general exhaustion forced her to give up flying. Other women tried, with less success, to emulate her multi-faceted career in aviation. They were to find, as Blanche Stuart Scott had done several years earlier, that the public looked to them to provide thrills and chills, but little else.

The unprecedented increase in aircraft production during the First World War threw a glut of planes on to the market in 1919. The Jenny (the Curtiss JN-4 biplane), for example, was widely available for sale at $600. Aviators of both sexes were quick to recognise its capacities for itinerant entertaining. The manoeuvrable Jenny could be landed in a small field and even sheltered in a barn, and 'barnstorming' began to sweep across America, as aeronautical performers flew from farm to airfield, demonstrating aerobatics, giving joyrides, making daring parachute descents and walking or dancing on the wings of their machines.

Although several male flyers who started out as barnstormers (including the young Charles Lindbergh) progressed to serious careers in aviation, women usually advanced no further. However, their skills were extraordinary, and intrepid. Lillian Boyer, a former waitress, perfected wing-walking and mid-air plane changes. Another waitress, Helen Lach, became a successful parachute jumper. Margie Hobbs was an enterprising circus trapeze artiste who changed her name appropriately to Ethel Dare and, as 'the Flying Witch', made rope-ladder transfers from one plane to another. Other aerial transferists used similar ladders to make rapid switches from automobiles to aircraft. Mabel Cody, who perfected this and many other perilous routines, ran her own Three Sky Ring Flying Circus. Phoebe Fairgrave, whose parachuting and stunting successes enabled her to start her own flying circus at the age of twenty, *did* manage to graduate from barnstorming to a more durable and serious job. She married one of her circus pilots, Vernon Omlie, and they eventually established their own flying school.

There were, of course, many fatalities before US government restrictions on low-level flying in 1929 drastically reduced these barnstorming programmes. The first American woman to be killed in an aeroplane accident was Julie Clark of Denver. She received her pilot's certificate on 19 May 1912 and tragically soon afterwards, on 17 June, died when her Curtiss biplane struck a tree at Springfield, Illinois. Amongst others to be killed were Gladys Roy, who had delighted her audiences by dancing the Charleston on her biplane's wings, and Bessie Coleman, the first black licensed female pilot, who was killed before she could save sufficient money from stunting to fulfil her ambition of starting a flying school.

Bessie was born in Texas in 1893. Although she did well at school, poverty prevented her from going to college. She established a chilli parlour in Chicago, and the venture was successful. However, she wanted a more fulfilling outlet for her talents and, after watching various barnstormers and stuntists in action, made up her mind to become a flyer. Apparently she was turned down by American training schools because of her race but, undeterred, spent all her chilli parlour profits in financing a trip to France where she obtained a pilot's licence. After returning to the USA in 1921 she announced that she would 'give a little colouring' to the world of aviation and eventually open her own flying school. Her aspirations were abruptly ended when, rehearsing for a Florida air show in 1926, she died after her plane crashed. It is some comfort to know that once she had become a successful barnstormer she found that profession 'free from prejudices' and was eager to bring other black people into it.

CHAPTER THREE

In Combat against the Kaiser

The Hun coming up behind shot at her continually . . . the tension of those fateful moments was terrible. Yet she summoned all her woman's pluck – the pluck that had come to the female sex in these days of war – and kept on flying.

THE insouciant last years of Edwardian ballooning and the expansive mood of mastering the elements that followed the Wright brothers' launching of their powered aeroplane can be seen in retrospect as the calm which preceded the ghastly storm of the First World War. Of course, developments in the aircraft industry were to proceed by leaps and bounds during those four years, and it must sometimes have seemed as if the terrible predictions of writers about war in the air were coming true. Some of these forecasts had even been made when ballooning was still a dream rather than a reality.

In 1737, forty-six years before Joseph and Etienne Montgolfier sent up their first man-carrying hot-air balloon, Thomas Gray had written a poem, *Luna Habitabilis*, which foresaw that

> The time will come when thou shalt lift thine eyes
> To watch a long-drawn battle in the skies.

In 1759, Samuel Johnson had recorded a warning note struck in a conversation with an artist:

> If all men were virtuous, I should with great alacrity teach them all to fly. But what would be the security of the good, if the bad could at pleasure invade them from the sky? Against

an army sailing through the clouds, neither walls nor mountains nor seas could afford any security.

Alfred, Lord Tennyson, also presaged aerial warfare in 'Locksley Hall' (1842):

For I dipt into the future, far as human eye could see,
Saw the vision of the world, and all the wonder that would be;
Saw the heavens fill with commerce, argosies of magic sails,
Pilots of the purple twilight, dropping down with costly bales;
Heard the heavens fill with shouting, and there rained a
    ghastly dew
From the nations' airy navies, grappling in the central blue.

Apprehension about the role of flying-machines in times of war rumbled on until, in 1909, Blériot's flight across the Channel caused several newspaper editors and writers to bemoan the loss of Britain's invulnerability. H. G. Wells, whose thrilling but disturbing novel *The War in the Air* had been published only a year earlier, declared that 'In spite of our fleet, England is no longer, from a military point of view, an inaccessible island.' Dolly Shepherd, the Edwardian Parachute Queen, wrote in her autobiography that when she decided to give up her aerial career in 1912 there was 'much talk of war . . . and it needed but little imagination to envisage the air as a battleground of the future. Somehow, some of the fun seemed to be going out of the skies.'

In 1914 thousands of young men, inspired by love of country and ideals of heroism, volunteered for the Royal Flying Corps (RFC) or the Royal Naval Air Service (RNAS). These were Britain's newest services; until 1912 war in the air had been the responsibility of the army. The Royal Engineers had flown balloons since 1876 and had maintained an Air Battalion since 1911. The RFC, founded in 1912, was originally administered by both the War Office and the Admiralty. The naval wing, however, broke away on the eve of the First World War to become the RNAS.

Before the war, despite the predictions of novelists and newspaper feature writers, British politicians had been slow to recognise the military capabilities of the aeroplane. In 1910 Lord Haldane had rather complacently remarked that, although the 'Englishman is usually behind' with new inventions 'like the submarine or the motor', in a few years 'He is generally leading . . . so, I suspect, it will prove to be with aircraft.' By 1911 a Defence Committee report suggested that time for Britain to establish leadership – or indeed even to stay in the race – might be rapidly running out. While France (let alone Germany, whatever might be going on there) had some 260 trained pilots in the services, Britain had 'about eleven . . . flying men in the army . . . and in the navy about eight'.

Obviously, in 1914, the recently formed RFC and RNAS were only too anxious to provide male would-be pilots with opportunities for training. For air-minded women, however, the doors remained firmly closed. Girls could help on a voluntary basis at RFC stations by cooking and scrubbing in their spare time. Some paid 'civilian subordinate' jobs, generally of a clerical nature, were also available to them. Such work was hardly in keeping with the *Per Ardua ad Astra* image. In fact, women and girls would have to wait almost until the end of the war before they could formally become part of a uniformed service formed specifically to back up male pilots and aircrews – and flying for females was still to remain taboo.

At the beginning of the war women pilots in Britain were few. Along with former balloonists, they reconciled themselves to helping the war effort in ways which made no use of their aeronautical skills. Dolly Shepherd became a driver–mechanic and served in France, while Gertrude Bacon joined the Red Cross. Hilda Hewlett, the first British woman to acquire a pilot's certificate, who had raced in the air against men since 1910, qualified on a Farman biplane at Brooklands on 29 August 1911. She established her own pilot-training school near London, and taught her son, then a Royal Navy sub-lieutenant, to fly. Evidently she wasted no time; he gained his certificate less than two months after she did. Possibly the only naval airman in the world to receive flying instruction from his mother, F. E. T. Hewlett was to become one of the early pilots of the RNAS.

France and Russia were apparently the only countries involved in the war where exceptions to the 'no wings for women' rule were made. In France Hélène Dutrieu, a professional trick cyclist who had become a pilot and been awarded the Légion d'Honneur, was the only female member of the Paris Air Guard, formed in 1914 to protect the city from possible attack by German planes or airships. There is, however, no record of her ever being on active service of this nature, and she died of old age in 1961 – unlike so many early pilots (thirty were killed in air accidents in 1910 alone). In Russia, Princess Eugenie Mikhailovna Shakhorskaya had received her pilot's certificate a few days before Hilda Hewlett, on 16 August 1911. As a result of her personal request to the Tsar, she became the first woman in the world to be appointed formally as a military pilot. She was posted to the First Field Air Squadron as a reconnaissance flyer, but, as with Hélène Dutrieu, no evidence exists of her going into combat. Princess Eugenie was not only strong-minded but resilient. Surviving both the war and the Russian Revolution, she seems to have switched allegiance from the old ruling classes to the Bolsheviks, and served in the Cheka (Bolshevik secret police) at Kiev in the post of chief executioner. This was surely, by any standard, strange work for a woman, and particularly for a princess.

The USA came into the war in 1917 and, as mentioned earlier, the record-beating pilot and stuntist, Ruth Law, tried hard but haplessly to get flying work with the armed forces, and had instead to harness her considerable expertise in aviation to make money for the Red Cross and Liberty Loan campaigns.

The Stinson sisters had also volunteered to fly with the services, but had been turned down. Nevertheless their contribution to the war effort was outstanding; even before America became involved in hostilities, the girls started to train aspiring pilots from Canada for service in Britain. Some thirty young Canadians passed through their flying school to become pilots in the RFC. Ironically, after America joined the Allies in 1917, the government banned civilian flying, and this brought about the demise of the Stinson school, which was already having problems as its small pool of intensively used planes was becoming depleted and disabled. (The Stinsons were not the only women who taught men to fly during the First World War. In England, Hilda Hewlett instructed fighter pilots, while in Germany Melli Beese, who had run her own Berlin flying school since 1912, instructed many fledgeling male pilots. An outstanding aviatrix, Melli's first attempts to gain a licence were sabotaged by some male colleagues who tampered with her plane's steering mechanism and partially emptied its fuel tank.

The Stinson sisters found new channels for their war work. Katherine went to France and Britain as an ambulance driver, while Marjorie decided to become a draughtswoman with the US navy's Aeronautical Division. Like Ruth Law, they also made fund-raising – and sometimes record-breaking – exhibition flights.

The frustration felt by Ruth Law and the Stinsons in not being allowed to fly with or for the armed forces, even in a supportive role, was echoed by other female flyers on both sides of the Atlantic; the formation of various women's service corps from 1917 onwards still brought about no relaxation of the 'no flights for females' rule.

In Britain the Women's Army Auxiliary Corps (WAAC) was founded in March 1917. The Women's Royal Naval Service (WRNS) followed in November of the same year, and in January 1918 the Women's Auxiliary Air Force Corps (WAAF) came into being. Girls in the newest service wore khaki, but before they could get used to their name and uniform, it was decided that the RFC and the RNAS should amalgamate into one new service to be called the Royal Air Force; this was in March 1918, and it was felt that the most recently formed female service should, to clarify its links with the male RAF, become known as the Women's Royal Air Force (WRAF). New blue uniforms were to be the order of the day, but many Wrafs found themselves wearing curious assortments of old WAAF-style khaki and air-force blue garb for some time to come, and for that early period of service often had to produce their own blouses, skirts and overalls. To add to the confusion, Waacs and Wrens who had been working on air stations were given the option of transfer to the WRAF, and frequent changes of leadership did nothing to inspire confidence in this new women's service. Unlike RAF personnel, Wrafs were enrolled, and not enlisted under the terms of military law. They were virtually civilians in uniform, handicapped in matters of leave and pay, sick care, injury compensation and unemployment cover. Nevertheless, despite all these difficulties, the new service got under way, and its members worked with enthusiasm, loyalty and dedication.

Recruits were known rather anomalously as 'airwomen'. Most of them never achieved the two-day course of initial training in service terms, discipline, drill and so on which regulations laid down, because reorganisation and shortages of accommodation meant that they were put to work straight away. Their quarters, too, often failed to come up to scratch; these were supposed to consist of rooms and furnishings of set size, number and type, but the curtains, couches, rugs, wash-stands, etc. provided were far less plentiful than they should have been. Inevitably there were grumbles about cramped, draughty and sometimes insanitary accommodation, but probably the most vociferous complaints from the early Wrafs focused on food. This ranged from being, at best, boring to, at worst, abysmal. According to Squadron Leader Beryl Escott's lively history of the service, *Women in Air Force Blue*, the caustic slogan 'Earwigs with everything' was justified.

Organisation at all levels was quick to improve, however, and airwomen were trained to become efficient cooks, mess orderlies, drivers and so on. Some RAF personnel, at first hostile to the influx of skirted aides, soon began to appreciate their worth. In 1918 Wrafs were employed in forty different trades, often surprising everybody with their newly developed technical skills. One of the main purposes of the WRAF was to provide female labour which would release aircraftmen – at first for other (presumably combatant) duties at home and abroad, and later, as the war drew to its close, for demobilisation.

Of course the signing of the Armistice in November 1918 did not mark an immediate sending home of service men.

Apart from the enormous administrative problems involved in such a vast procedure, an occupation force in Germany was considered imperative, and WRAF domestics, clerks, telegraphists, store-keepers and drivers formed part of this Watch on the Rhine. They acquitted themselves well and were complimented by high-ranking RAF officers and government officials for their efficiency and high standards of discipline. Nevertheless, despite all the individual, group and organisational skills which the WRAF had developed, its disbanding was begun in November 1918 and completed by the spring of 1920. Many women felt that the service had been unnecessarily short-lived, and that it should have been retained as a peacetime body. At its greatest strength at the end of 1918 it had numbered almost 25,000 women.

Although many WRAF volunteers had, so to speak, rubbed shoulders with aircraft, and helped with their maintenance as welders, sheet-metal workers, turners, riggers, vulcanisers, carpenters and fitters, their opportunities for flight were strictly unofficial, occurring only when they were taken up, informally, as passengers of RAF pilots.

References to members of the WRAF rarely crop up in the fiction of the First World War, partly, of course, because of its late formation. (Women's magazines of the period were similarly reticent about the most recent of the men's services, the RAF. In *Forget-Me-Not*, for example, cover after cover offered glamorised illustrations of naval and army officers, ratings and tommies, while flyers didn't get a look-in. RAF images in 1918 seemed far less vivid in female imagination than the silver-winged ones of the Second World War.)

In stories for men and boys the focus was quite different. Flyers were heroes; the principles of aerodynamics were intriguing; planes were lovingly and meticulously illustrated; air battles were thrillingly described. First World War flying stories were popular not only between 1914 and 1918, but also during the 1920s and 1930s. Naturally they were superseded during the 1940s and 1950s by tales which exploited the achievements of the pilots of Hitler's war. However, by the 1960s stories of skirmishes between Sopwiths and Fokkers were inching back into favour, and sometimes edging out from boys' books and papers the sagas of Battle of Britain dog-fights between Spitfires and Messerschmitts.

Throughout 1916 readers of the monthly *Boy's Own Paper* were being treated to several features on aviation. These included finely detailed, full-colour plates of Allied and German *Aeroplanes and Airships* and *A Duel in the Air* ('How Flight Sub-Lieutenant R. A. J. Warneford V.C. Destroyed a Zeppelin over Ghent') as well as a series of stories celebrating 'the heroic exploits of British, French, Belgian, Russian and Italian Aviators at the Front'. Over the same period, girl readers of *Forget-Me-Not* were served only one flying feature – a story by Violet Methley called 'Snapdragon' which, like Kitty Smart's adventures in the 1911 *Sunday Stories*, is an account of an elopement by air. (British writers for women and girls firmly favoured this theme and the romantic use of aeroplanes, even at a time when they were being used so dauntingly as instruments of death and destruction.)

'Snapdragon' contains no reference to the war, although the fact that Gillian – who helps the lovers – is referred to as 'a lady gardener' suggests that she is engaged in horticulture as her voluntary war work. Gillian feels protective towards Drusilla Deane, 'a graceful, fair, little [and apparently leisured] creature, with appealing grey eyes' whose 'exceeding prettiness is rather spoilt by an air of wistful anxiety'. This clouding of her beauty has been brought about by the bullying of an ambitious mother, who urges Drusilla to respond to the romantic overtures of Mr Harper Enson. He is a rich American – and we are reminded that this is 1916, when the USA had still not decided to come into the war and was subjected to some resentment in Britain because of this. Enson is described as overweight and overbearing 'with a loud, vulgar laugh, and a louder check suit'. Even Gillian, who comes across as a generally well-balanced character, 'falls in hate with him at first sight'.

Gillian realises that Drusilla has a more suitable – though impoverished – upright young English lover, in the shape of Jack Trelawney, who leaves notes for his sweetheart in the mouths of some large snapdragons (antirrhinums), and literally whisks her off in his biplane from under the nose of the 'utterly dumbfounded American'.

First World War popular fiction generally reflected reality by casting its war-working heroines in the roles of Voluntary Aid Detachment workers, munitions-makers, land girls and ambulance drivers. There really were no female flyers who could be used as role models. The height of ambition for Angela Brazil's schoolgirls, who were bursting to do their bit to show the 'Huns' what it meant to be British, was to dash off to the front 'to nurse the wounded, drive a transport wagon, act as secretary to a staff officer, or even be a telephone operator'. The weekly *Penny Pictorial* was designed for women rather than girls, and in 1916 it ran 'A Series of Romances Resulting from the Vast Changes in the Spheres of Women'. These were all written by an author called E. Almaz Stout; they followed the formula of the story starting with incredulous male disbelief that a slip of a girl could cope with driving a van or being a bank cashier, until the heroine pluckily proves her worth by foiling spies and other baddies before deciding to abandon war work for romantic fulfilment: 'With a little moaning sob, Molly fell forward into Harry Cardew's arms. But she was made of British fibre, and she knew she must not faint or give way yet.'

A lot of the more serious fiction about the First World War was written retrospectively by men, and, as Vera Brittain pointed out during the early 1930s, little of this told the story of women's involvement in war work. Evadne Price, writing as Helen Zenna Smith in *Not So Quiet . . .* (1930), attempted, with over-dramatic clout, to give a real rather than a glamorised account of the life of a front-line female ambulance driver. Enid Bagnold, Vera Brittain and others provided authentic accounts of military nursing experiences, but the WAAC, WRNS and WRAF had no dedicated chroniclers.

While the war was still very much under way, it was left to William Le Queux, a master of popular spy and sleuthing fiction, to spotlight the female flyer. His 1917 *Beryl of the Biplane* ('Being the Romance of an Air-woman of To-day'), was reprinted in 1918 and 1919. Its episodic nature, with every chapter telling a separate, spy- or Hun-catching story, suggests that Beryl's aeronautical counter-espionage activities might originally have enlivened the pages of a pulp weekly for women.

Beryl Gaselee, 'the intrepid woman aviator', is not only 'very pretty' but 'devilishly clever'. (To make any kind of career in aviation then, of course, she would have to be.) The first chapter tells us that Beryl has flown the Channel 'three times alone' and on four occasions with her fiancé, Ronald Pryor ('Ronnie'). Beryl's 'wonderful nerve for a woman' is frequently described; she stunts with panache, flies 'as well as any RFC Squadron Commander', does double war duty by acting as a 'uniformed chauffeuse' when on the ground, goes up night after night with Ronnie in his big biplane to hunt Zeppelins, shoots down Fokker pilots and bombs enemy installations in occupied Belgium. Nevertheless there is nothing butch about our Beryl, and when she rips off her 'workmanlike, windproof overalls', leather helmet, goggles and gauntlet gloves, 'she presents a charming figure of that feminine type that is so purely English', with fair and fluffy hair and 'big, wide-open blue eyes aglow with the pleasures of living'. She is so dainty and *petite* that fellow airmen are amazed that she can manipulate her fiancé's powerful plane, and 'the way in which she manipulates the joystick often, indeed, astonishes Ronnie himself'.

With so much going for her, it is not perhaps surprising that Beryl's only domestic duty appears to be the daily brushing of 'her little black pom'. She is not a member of any official service or organisation but, like Ronnie, seems to hunt down on a freelance basis 'those who drink the health of the Kaiser in champagne'. (William Le Queux's characters – even his hated Huns – do everything with style.) In the air Beryl can use wireless, bash out Morse code messages with her searchlight and easily manipulate a heavy machine-gun. The only time she almost loses her cool is when her lover lies slumped against her in the cockpit, apparently bleeding to death: 'Gad, Beryl . . . they've got me – the brutes!' As she has to pilot the plane back to base she can no longer use the machine-gun: 'Their lives now depended upon her skill in manipulating the machine.' She banks, dives and manages to evade the 'angry spurts of red fire' from the enemy plane. Once safely home, she becomes Ronnie's nurse – an expert one, of course – and, when he is well enough to walk down the aisle on crutches, his wife.

Beryl of the Biplane is tremendously entertaining, although possibly not in quite the way Le Queux intended. It is plentifully adorned with exciting aeronautical flourishes, but one suspects that the author did not know a great deal about the techniques of flying, although he was attracted to the romance of it. One thing that is particularly surprising to the reader in all Beryl's 'deadly air battles' and very long reconnaissance flights is that she never has to worry about refuelling the biplane. It stays up for hours, apparently, simply by drawing power from the enormous energy and determination of this one female flyer who *did* go into combat against the Kaiser.

CHAPTER FOUR

Aristocratic Ladies and Anthropomorphic Animals

Illustration from Mary Tourtel's first Rupert Bear strip 'Little Lost Bear', *Daily Express*, 1920.

Only for this very brief period was the human race to enjoy the freedom of the air with the birds . . . one could fly over frontiers and instead of being looked upon as a suspicious character one was acclaimed and welcomed as a hero or a harmless lunatic.

LINDA RHODES-MOORHOUSE,
Kaleidoscope (1960)

LINDA MORRITT was born in 1886 in the North Riding of Yorkshire. Her home, Rokeby, was stately and splendid and run by a positive army of domestics. With an enlightened father who valued 'his children's freedom of spirit before their physical safety', she was able to make the most of 'the life of leisured elegance lived by the lucky few' by 'running wild' in the wonderful but rather dangerous countryside around Rokeby. The 'magical' experience of exploring and negotiating the 'swiftly running rivers with their steep and rugged cliffs' instilled in the small child a taste for freedom and adventure which was later to find an outlet in flying.

Then came hunt balls, the round of visits to other country houses, the London season, a visit to Cairo and a short spell at the Augustus John Art School with her close and lifelong friend, Anne Rhodes-Moorhouse, whose brother Will was to become Linda's husband. Linda was immediately attracted to this dynamic 'Will o' the Wisp', who was never still and always involved in some project or other arising from his 'consuming passion' for engines and speed. He was addicted first to motor-bikes, then to racing cars and, finally, to aircraft. In 1911 he manufactured and flew a high-wing monoplane of his own design (the Radley-Moorhouse) after suffering many accidents in testing it. Not surprisingly, Will persuaded Linda that the best possible honeymoon would be a flying one. At that time – 1912 – the Channel had, of course, been flown solo by Blériot and others; it had also been crossed by pairs such as Gustav Hamel and Miss Trehawke Davies. However, Will decided that it would be 'great fun' to take the first threesome – himself, Linda and a reporter, J. H. Ledeboer – across the celebrated 21-mile strip of water. After spending their wedding night in the elegance of the Savoy hotel, Linda and Will travelled to Douai, north of Paris, to collect and test the Bréguet biplane which was being prepared for the Channel flight. It was far from ready, so, as well as spending her honeymoon nights in a very scruffy French hotel, Linda spent all the days of it with Will at the draughty aerodrome where he was working with the mechanics on the biplane. It was, as she commented later, a strange honeymoon but, culminating in a successful (though crash-landed) flight from France to England, it drew the couple more closely together than a conventional, luxurious one might have done.

At this time Linda was content to remain Will's passenger; her decision to become a pilot in her own right came much later. Their son, William, was born in 1914 – so quickly, despite his 9¾ lbs, that there was no time to fetch the doctor, and the redoubtable Will had to help the baby into the world. Sadly there was to be little subsequent relationship between father and son; Will joined the RFC immediately on the outbreak of the First World War, and was killed after a single-handed raid on a big railway junction and marshalling yard about 35 miles behind the German lines. For his gallantry in this action he was awarded the first VC to be bestowed upon a flyer.

Linda accepted his death after their short-lived marriage with fortitude, and devoted herself to her son William, who resembled his father in many of his tastes and interests. While still a schoolboy, he was fired with an ambition to fly when Adelaide Cleaver, the mother of one of his friends, took him for a flight in her Percival Gull. Despite certain apprehensions, Linda agreed to finance his flying lessons, and, by selling a valuable stamp collection bequeathed by her mother-in-law, bought him a Gipsy Moth. William became known as the 'Flying Etonian' when he acquired his licence at the age of seventeen.

Once again Linda found herself passenger to a skilled and enterprising pilot. She accompanied him on flights abroad and on cross-country trips to stay with friends in England, and was soon inspired to obtain her own pilot's licence – even though 'forty-five is not the ideal age to start flying'.

Her love of being in the air never quite matched the total joy which both her husband and her son found in flying. She comments in *Kaleidoscope* that 'besides these moments of intense pleasure there were hideous ones of fear and appre-hension.' Her anxiety was always for William rather than herself, and, tragically, her worst fears were realised when William, after volunteering for the RAF at the beginning of the Second World War, and being awarded the DFC, was shot down in his Hurricane and killed during the Battle of Britain in 1940.

After William's death Linda went to live at Mortham, a beautiful old peel tower near Rokeby, and once again had to build a new life for herself after bereavement. Her 'restless spirit' took her away on many long and adventurous jour-neys, but after the ending of the Second World War, her days as a pilot were over. Nevertheless, like her husband and son, she always seemed to be in her true element in the air. Reflecting on her flying experiences towards the end of her long life, in the 1960s, she commented:

> The earth was so far away it hardly seemed to exist and to be an enemy rather than a friend. Of course one is always alone in the citadel of oneself, but when one is alone in the sky one seems to realise this fully for the first time. I think it helps one in ordinary life to have experienced this. Perhaps it makes one more self-reliant.

Linda Rhodes-Moorhouse must surely have been the least eccentric of the clutch of upper-crust English ladies who took up flying in the late 1920s and early 1930s. One of the most

colourful of these was Lady Heath who, as Sophie Mary
Pierce Evans, was born in County Limerick in 1896. Her
childhood, unlike Linda's, was rather restrictive, and partly
because of this, her natural high spirits made her always
willing to accept 'a dare' or a challenge. Aviation and campaigning for women's rights were to absorb her in adult life;
she had attained a Dublin science degree before starting to
fly at the age of twenty-two, but as a young girl, work and
study had taken second place to 'lots of scrapes and esca-
pades'. Writing about her childhood in the magazine
School-Days (1 December 1928), she explained that the aunts
with whom she lived 'did not approve of sports for girls' and
constantly moved her from one school to another, because
hockey or athletics figured too prominently on the timetable:
'I was taken away from one school . . . on account of the fact
that not only was hockey played there, but one match in the
term was played against a boys' school! I was at once transferred to another school where the girls went out, two and
two, every day for a nice long walk in galoshes!'

Her outlet from her aunts' excessive concern was to risk
punishment or expulsion in 'wild' and anti-social behaviour.
At boarding-school she discovered a secret passage and hid
mutton bones there together with a message in 'blood' (red
ink), supposedly from a prisoner who had been tortured to
death; she hoodwinked a headmistress by pretending to be
the aunt of a pupil and taking her out to tea; and she climbed
from her dormitory into the grounds of another school to
appropriate some of the refreshments laid out there for a
garden party. Her taste for the intrepid stood her in good
stead later on, when visiting outlandish places on her long-
distance flights and battling with reactionary officialdom.

She married an older man, fairly soon became a widow,
and, as Mrs Elliott-Lynn, became a well-known pilot and
campaigner. (Her second marriage was later to give her a
title and the financial means to widen the scope of her
aeronautical career.) At first she had to earn her living, and
to do this as a pilot a commercial licence was necessary. Her
first major struggle was with the International Commission
for Air Navigation, who had decreed in 1924 that would-be

commercial pilots must be of the male sex. She challenged the Commission on the grounds that she knew more about women and their capabilities than the doctors (all male) who had imposed the ban. Her qualifications to make such a claim were impressive; as well as being an experienced pilot, she had a degree in physiology, was a champion high-jumper, and the founder of the Women's Amateur Athletic Association of Great Britain. Nevertheless she had to suffer the indignity of proving to the Commission that she could function efficiently as a pilot at all times of the month! Nothing daunted, she did so. By May 1926 the ban on female pilots carrying passengers was rescinded.

Sophie – then still Mrs Elliott-Lynn – decided to demonstrate that women could fly as well as or even better than men. In 1927 she set an altitude record of 16,000 feet, entered and won several air races, and gave lectures on aviation in various parts of the country. It is no wonder that the magazine *Flight* recorded that 'Mrs Elliott-Lynn has perhaps done more for her sex [in aviation] than any other woman.' However, her ambitions surged even higher, and, fairly soon after marrying Sir James Heath, she planned to make the first solo flight by a woman from South Africa to England. The first trip in the reverse direction had been made by Flight Lieutenant R. Bentley of the South African air force only a few weeks before Lady Heath set off from Cape Town to Croydon. Bentley's trip took twenty-seven days, while hers was to extend over three months, from 12 February to 17 May 1928.

The difficulties which beset her included the loss of large-scale maps of the route, extremely severe sunstroke, and the fact that she was not allowed to fly over at least two sections of the journey without being 'chaperoned'. Bentley obligingly provided protective escort over both the Sudan and the Mediterranean. A further hazard, of which Sophie was unaware at the time, was that her plane – an Avro Avian III G-EBUG – had been shot at when she was flying over the North African coast, and holed in one of its wings.

Lady Heath continued to campaign for opportunities for women in aviation and to provide inspiration for many aspiring flyers. She was generally fearless, drawing security from the fact that she maintained her machine carefully: 'I did the tappet clearances every day, no matter how short the flight was, and cleaned the petrol and oil filters.' The only thing which really worried her was the possibility of crashing into the sea; on her Cape Town to Croydon flight she wore two inflated rubber tyres which she hoped would act as a life-jacket, but these burst at 7000 feet and fell into shreds around her neck.

She believed in being well prepared for every situation, and in addition to essential medical supplies, she took on her long trips plentiful reading matter and almost every kind of outfit from full evening dress to tennis gear. It is not surprising that at the end of each flight she generally managed to climb from her plane looking immaculate. She is supposed to have claimed that aviation was safe enough to permit a woman to fly over Africa 'wearing a Parisian frock and keeping her nose powdered all the way'. Lady Heath's courage and determination caused her to be elected in the USA, Lady Champion Aviator of the World. In Britain, however, she received brickbats as well as as bouquets, and was sometimes dubbed Lady Hell-of-a-Din because of her flair for self-advertisement.

Her friend and rival, Mary, Lady Bailey, vied with her in setting speed and altitude records. Both had Irish backgrounds, tremendous enthusiasm for aviation, and husbands who were wealthy enough to indulge this. (When, for example, Lady Bailey's machine became a write-off in Tanganyika, she simply cabled her South African millionaire spouse for another one – which reached her after only a few days.) After becoming the first woman to fly across the Irish Sea, she decided to tackle the flight from London to Cape Town in a de Havilland Moth. Subjected to many delays, she eventually took off in early March 1928, arriving on 30 April. After spending a few months in South Africa, she made up her mind to attempt the return trip to London. Again there were hold-ups and mishaps, and the flight occupied the period from 21 September 1928 to 16 January 1929. Nevertheless she was the first woman to make this solo return trip.

Unlike Lady Heath, Lady Bailey was capable of travelling light, being unconcerned about sartorial elegance. (At a Khartoum reception where both ladies were guests of honour Sophie characteristically sported the splendours of evening dress while Mary appeared in a tweed flying suit.) And despite her enthusiasm, Lady Bailey was inclined to talk about aviation in a casual, almost vague, way.

Mary Petre, who became the Hon. Mrs Victor Bruce after marrying a racing driver, had an even more relaxed approach. Attracted to a Blackburn Bluebird which she saw on display in London, she drew a round-the-world line on an atlas and asked the AA for the appropriate air-route maps. Someone within the organisation asked her when she had learned to fly, and was disconcerted to receive her unruffled reply: 'I haven't yet, but I will before I go.' She did too – and in only a couple of weeks.

Her planned 1930 trip, from England across Europe and Turkey to the Middle East, then to India, Burma, Hong Kong, Shanghai, Tokyo and America, seemed an impossible venture for so inexperienced a pilot. She sensibly decided to cross both the Pacific and the Atlantic by ship. Nevertheless her five-month journey involved horrible hazards, including forced landings in the Syrian desert, on a quicksand in the Persian Gulf and over the Potomac River in the USA. She and her plane had to struggle through sandstorms, an attack by vultures, a tropical rainstorm, and, of course, the inevitable fog over the Channel and England. Highspots of her circumnavigation of the globe were clearing the crowd from an Ankara stadium by dropping smoke bombs (she had missed the aerodrome and had to land somewhere), a motor-cycle police escort in San Francisco, and being met by Amy Johnson and a gaggle of Blackburn Bluebirds, which escorted her from Lympne to Croydon at the end of the long, long trip which had taken in some twenty-three countries in four continents.

Towards the end of this truly remarkable journey, Mary Bruce wished it was all over as she was 'so tired of flying, flying, flying'. Nevertheless her resilience soon returned, and she was one of the few British women of her day who managed to pursue a serious career in aviation. After a spell in the British Hospitals' Air Pageant Flying Circus she operated a company called Air Dispatch. This carried not only freight but passengers, and it included an air ambulance service. Air Dispatch went from strength to strength, becoming the fastest air service between London and Paris.

The Hon. Mrs Victor Bruce's interest in speed extended beyond aeroplanes. An excellent horsewoman, she obtained first and second places at various national events in 1938. Much later, and well into her seventies, after driving a Ford Ghia Caprice round Thruxton race track at 110 miles an hour, she is reported to have said, 'Going slow always made me feel tired.'

Her energies spilled over into writing. As well as two accounts of her flying achievements she produced a series of short stories for the *Sketch*, which were collected and published in book form in 1930 under the title of *The Peregrinations of Penelope*.

The eponymous, upper-crust heroine stars in every sequence and, like her creator, is frequently on the move. A resolutely bright young thing, she drives 'a dinky little two-seater' car, breaks the speed limit all over London and the English countryside and drives recklessly in Paris. She rides with the hunt in two counties, motor races over water in Wales and on the track at Goodwood. Expectedly, with her predilection for speed and excitement, she also takes to the air. Echoing her author's real life experience she quickly and almost nonchalantly becomes a confident aviatrix saying firmly to her beau (an unnamed character who is involved in all her exploits and narrates the stories), 'I said we're going to fly. And so we are. I've been taking lessons for a week or so on the quiet, so that you shouldn't put me off, and I qualified for my pilot's certificate yesterday.' She has also (with the help of her affluent, indulgent father) bought 'the dinkiest little Flutterby' in which she insists on bearing her male admirer aloft. The flight turns out to be hair-raising for him. Penelope performs various stunts – apparently for the first time. She loops the loop and 'side-slips and side-slips and side-slips' in what seems to her reluctant passenger to be a 'dive to death' before she eventually lands ('I'm not very good at it') in a field of sheep, where, unruffled, she 'proceeds to powder her nose in the driving mirror'.

Surprisingly, Mary Bruce does not provide us with any real insights into the challenges of aviation, and her heroine never flies again. The mood of her facetiously chronicled adventures – like that of so much light fiction of the 1920s – is Wodehouse *manqué*, but Penelope's saga is considerably enhanced by the black and white line illustrations of Joyce Dennys, which stylishly convey the atmosphere of the period.

Another impressive flying aristocrat was the Duchess of Bedford. Born Mary du Caurroy Tribe, she married Lord Herbrand Russell in 1888, five years before he inherited his title and the vast acres of Woburn Abbey. For many days in each year Mary sought quietness and solitude by going out to sea in her private yacht, or on bird-watching expeditions. Having scant interest in domestic affairs, or even in her young son, she dedicated the bulk of her time to a cottage hospital which she founded at Woburn in 1898, and which became a military hospital between 1914 and 1918. It was only in the last ten years of her life that she decided to take up aviation. Over sixty and with a private income of £30,000 a year (a fortune in the 1920s), she was able to own several aircraft at once and to build a hangar for them on the Woburn estate.

The 11th Duke of Bedford appears to have been long-suffering. Having supported her hospital for several decades, he now had to put up with his wife going away, sometimes for long periods, on her various flights – and with her various hired pilots. The Duchess obtained a pilot's licence but, on the whole, preferred being chauffeured through the skies to taking the controls. As a kind of consolation prize, she tried to spend a half hour with her husband every evening when she was not away on a long-distance flight. Generally they took a walk together on these occasions, and even on her typically busy days (nipping over in her Gipsy Moth to Brussels for a couple of art exhibitions, and then back to Reading) she liked, according to her diary entries, to round off 'a varied and well-filled day' with that companionable evening stroll at Woburn.

During the period of her enthusiasm for aviation, she liked to employ two pilots. Captain D. C. Barnard and Flight Lieutenant J. B. Allen both performed the double duties of aerial chauffeur and flight instructor. The Duchess was looping the loop only a week after Barnard began to give her the first of the lessons which enabled her eventually to acquire her licence, and Allen completed the instruction course. Despite the enforced intimacy of close cockpit proximity and the sharing of vicissitudes on long-distance flights, Barnard and Allen, even when rapping out instructions to the Duchess during a minor emergency in the air, never failed to address her as 'Your Grace'.

She was a much-respected though rather remote character. She suffered from deafness brought on by typhoid at the age of sixteen, and this, compounded with sometimes unbearable head noises, isolated her from a great deal of human contact. Apparently flying provided release from the buzzing in her ears that was almost constant when she was on *terra firma*.

The Duchess's most celebrated flight took place before she had obtained her pilot's licence. With Barnard at the controls of her Fokker F.VII.A plane, she proved in 1929 that it was possible to fly – or be flown – to India and back in eight days 'without undue fatigue'. On this trip the dauntless Duchess had to spend a great deal of her time aloft pumping petrol, but afterwards affirmed that she had enjoyed it, and admired the panoramas unfolding above, below and around her.

In the following year she and Barnard went to the Cape and back. The Duchess was now able to share the piloting, and on the whole they had a smooth trip, though it was spiced by an emergency landing – when an oil leak was suspected – somewhere near Lake Victoria; the Duchess was received with enthusiasm by a 'weird collection of natives dressed in girdles of beads only'.

In 1933 the Duchess went with Flight Lieutenant Allen on a month's trip to Europe and the Middle East. He was her favourite pilot and when, only months afterwards, he was killed in a Cairo air race, she felt that she had lost a great friend. Her relationship with her next pilot, Flight Lieutenant Preston, was prickly, as he appeared, in her view, to lay down the law rather than to advise. However, they got on well enough to make a flight together across the Sahara which lasted for six weeks. This included several interesting excursions on the ground, such as the Duchess's informal inspection of an Emir's harem.

Her presence graced many flying meets and provided both publicity and social acceptability for female aviation. Her seventieth birthday was marked by a brief solo flight to London but, soon after this, declining health prevented her from making further aerial trips. With failing eyesight and increased deafness and head noises, she became more and more depressed. It began to seem unlikely that she would pass the next annual medical test which the retention of her pilot's licence demanded. To add to her sense of frustration, the Duke informed her that he would no longer be able to afford to cover the costs of her beloved hospital. In March 1937 she took off on a solo flight and never returned. Ostensibly she had taken to the air to bring her solo flying time up to 200 hours; however, when parts of her plane were later washed up from the sea – way off the course she had said she would take – it was suspected that the Duchess had deliberately ended her life.

Celebrated for very different reasons, Mary Tourtel, the popular book illustrator, was an early flying enthusiast. Even before she created the endearing and perpetually resilient Rupert Bear for the *Daily Express* in 1920, she had excelled in sketching animals, often in natural country settings, and part of her joy in flying came from 'seeing the land as the birds saw it'. She and her husband Herbert Tourtel – the *Express*'s night news editor – were among the passengers in the 1919 Handley-Page biplane record-breaking trip from Hounslow Heath to Brussels. This flight, which took two hours and thirty-five minutes, marked the beginning of regular passenger services from Croydon to a number of European capitals.

Mary lost no time in projecting her passion for flight upon her furry hero. Rupert is hoisted twice in his initial adventure ('Little Lost Bear', 1920), first by a bunch of balloons which he is clutching, next by an airship. Some of the airship pictures show the rustic environs of Rupert's village, Nutwood, spread out below, rather as Mary would have seen the countryside from her air trips in the Handley-Page and similar machines. After his first flying exploit, Rupert has only to look at a laundry basket to find himself being transported through the skies in it, while Mary also air-lifts him by means of magical bicycles or shoes, giant birds, umbrellas, the backs of dragons, kites, clouds and, of course, a variety of conventional aeroplanes. (Mary's successors, when she had to abandon the strips in 1935 because of failing eyesight, maintained the aeronautical tradition. Alfred Bestall in particular proved extremely inventive about Rupert's modes of becoming airborne. In the 1980s, these included pogo sticks and flying saucers.)

Rupert was of course a boy, albeit a beary one, and his closest chums – Willie Mouse, Edward Trunk, Bill Badger, Podgy Pig and Algy Pup – were also male. Though not air-lifted quite so often as Rupert, they took to the skies fairly frequently. Small girl readers of Mary Tourtel's and Alfred Bestall's addictive picture-stories, however, might have been surprised that Rupert's female associates seemed far less likely to fly. Of course there were far fewer girls than boys in the saga – but knowing of Mary Tourtel's interest in flight one might have expected her to allow Rupert's feminine friends, Margot and Barbara, and more of the little princesses whom he saved from dragons and corrupt pretender-kings, to enjoy the freedom of the air more often.

Rupert and Co. were not the first anthropomorphic animals to become cult comic-strip figures, but they seem set to be the longest-lasting. They were originally inspired by Teddy Tail of the *Daily Mail*, the cheery, knowing little mouse who sprang to favour with the public from his first appearance in the paper in April 1915. Teddy was originated by the notable book illustrator Charles Folkard, and, in the battle for increased circulations, several other papers tried to emulate his success by creating their own furred or feathered heroes and heroines. Teddy's chums included a girl, Kitty Puss; the *Daily Mirror*'s strangely assorted trio of Pip, Squeak and Wilfred (a dog, a penguin and a rabbit) had its statutory female in the shape of Squeak, the motherly, hand-bag-clutching and domesticated penguin. Bobby Bear in the *Daily Herald* had two female friends, Maisie Mouse and Ruby Rabbit. All these groups of beaming, bouncing and essentially engaging animals – girls as well as boys – took to the air from time to time, though not with the same frequency as Rupert did.

In 1925 one of the most talented artists in British comics, Herbert S. Foxwell, created Tiger Tilly and Co., the animal girls of Mrs Hippo's boarding-school, for the weekly *Playbox*. Tilly's coterie included a bear, a monkey, an elephant, an ostrich, a kangaroo, a dog and a parrot. Her rival and enemy was Pearl Porky, a pig. The girls were in fact skirted versions of a group of male animal characters, Tiger Tim and Co. who had been popular in papers and comics since 1904 and continued to appear fairly regularly until the late 1980s. Tilly and her chums were particularly addicted to airborne adventures; until they faded away soon after the end of the Second World War they spent a lot of time in aeroplanes (which bore no resemblance to any known machines).

Apart from these animal picture-strips in comics and annuals, stories about girl flyers were fairly rare during the 1920s, in both children's and adult publications. The heyday of the 1930s, in both fact and fiction, was still to come. One of the first text stories for air-minded little girls was 'Little Miss Aeroplane' in *The Sunbeam* of 2 December 1922. Significantly, Muriel, who longs to fly, has to pretend to be a boy before she can get a trip in an aeroplane. Her normally indulgent father vehemently vetoes her hopes of going up in 'a flying-machine of any kind whatever; neither balloon, nor aeroplane nor seaplane nor anything else that may be invented'. To his offspring he does not explain why, although readers are subsequently told that it is because he has seen 'a terrible accident happen to his dearest friend through an aeroplane going wrong', and has since nurtured a horror of flying.

Muriel, however, is determined to get into the skies by fair means or foul. She haunts the local aerodrome and becomes friendly with Charlie, a cabin-boy on what is called a cross-Channel liner (but which looks like a fairly small monoplane in the story's illustration). With Charlie's connivance, Muriel slips behind a hedge and changes into his 'smart, navy blue uniform, with bright buttons that shone like silver'. She tucks her frothy curls into his cap and maintains the illusion that she is a boy until, thrown off course by fog, the captain lands far from the local aerodrome on the return flight, and Muriel, realising that she will be late home, bursts into tears. The male pilot takes pity on her and gets her home before nightfall to face her father who is 'so glad to see his disobedient daughter home' that he 'forgets to be cross'.

The implication is that an ambition to fly is unnatural in a girl and must therefore result in some kind of retribution. (Poor little Muriel, by the way, looks as if she is only about eight years old.) A comic for slightly older children, *Puck*, struck a more positive note a few months later in October 1923, with a serial called 'The Flying Girl'. Mind you, its heroine Jo Grant also has to pretend to be a boy in order to get a flying job. However, it is acknowledged that she is as competent a pilot as her brother, Dick, and she copes with frenzied storms, forced landings and fearful baddies as gamely as anyone. Jo seems to be the first teenage British girl flyer cast in a similar mould to the American girl aviators whose golden wings had taken them up a decade earlier. By the end of the 1920s, after the impact of the aeronautical achievements of Lady Heath and Co. had filtered into fiction, progressive notes were beginning to be struck in books and comics for girls. The monthly magazine *School-Days*, which had lauded the exploits of the real-life upper-class aviatrices, produced a short story called 'Sky High' by June Grey in its 26 October 1929 issue. Schoolgirl twins Daphne and Dilys (Daffy and Dilly) are air-crazy, although they have not yet had the opportunity of going up in a plane. A small joy-riding set-up arrives near their boarding-school and, after deviously getting permission to go into town on the pretext of buying a birthday present for their mother, the twins exuberantly go up on a 10s-6d twenty-minute flight.

The jinx that so often comes into play when schoolgirls try to deceive their headmistresses produces engine trouble – and the pilot has to land his plane on (guess what?) the hockey pitch of Daffy's and Dilly's school where the annual Teachers-versus-Sixth-Form match is taking place, watched by a large crowd of girls and staff. Fortunately the twins' identities are concealed by the overalls, helmets and goggles which they have had to wear for the flight, and the pilot, who does *not* believe that would-be girl flyers deserve to be punished, helps them to avoid recognition, and retribution from the Head for their illicit trip to town.

In the same year *The Empire Annual for Girls* published 'A Thrilling Adventure of the Air' entitled 'Flying for the Mounted'. Set in the wastes of Canada, and written by William Macmillan, who seemed really knowledgeable about flying, this story makes an appropriate prelude to the 1930s' flood of popular flying fiction for women and girls of almost all ages, which was soon to appear. Eighteen-year-old Beth is a fully qualified pilot who, working for her family's small airline, flies the mail to a group of isolated and ice-bound islands. On one of her trips she observes a man fleeing for his life from a dozen pursuers, and niftily lands and rescues him. It turns out that he is a Mountie, and not only does Beth save his life but, ultimately, helps him to 'get his man'. Fictional girl flyers were at last beginning to become accepted in realistic tales as well as in fantasy exploits of furry creatures.

CHAPTER FIVE

American Legends and Luminaries

Russell Keaton's Flyin' Jenny, 1941.

If enough of us keep trying, we'll get some-place.

AMELIA EARHART (1929)

WITH the resumption of civil flying after the ending of the First World War, America was once again in the forefront of aviational energy, expertise and achievement. Stunting remained as a hangover from barnstorming, and more and more flyers – including women – were being used for aviational advertising. During the 1920s aircraft manufacturers were at pains to point out to the public that flying was safe. 'Airmindedness' became a promotional term: attractive women beamingly adorned the pages of magazines publicising the trappings of aviation from goggles to flying-suits. The implication that flying could be indulged in by girls as well as men emphasised its safeness and accessibility.

Most of all, feminine involvement in aviation was demonstrated by races and competitions, although it was not until the end of the decade that women's records were classed and publicised as such. Before then their achievements were lumped under the heading of 'Miscellaneous'. The first Women's Air Derby in August 1929 marked a more serious attitude towards female flyers. It was arranged as one of the attractions of Cleveland's National Air Races, and it was to become an annual and extremely prestigious event. It gave women not only the chance to establish a reputation in the flying world but the opportunity to win its $2500 prize (a considerable amount of money at the end of the 1920s). The Women's Air Derby, which was to become generally known as the Powder Puff Derby, was no token event. It began near Los Angeles, and followed a 2800-mile route to Cleveland,

87

with the expected duration of a week. A history of the Powder Puff Derby could fill a book by itself; a report of its first year alone would include many colourful facts and features.

To qualify for the race it was necessary to have achieved 100 hours of solo flying time. Foreign as well as American pilots could take part, and the first Derby attracted Thea Rasche from Germany, who became a famous stunt pilot in her own country, and Jessie Miller from Australia, who had co-piloted a London to Darwin flight. The American competitors included at least three women who nurtured ambitions to become the first female to fly solo across the Atlantic; they were Amelia Earhart, Ruth Elder and Ruth Nichols. In 1928 Amelia had officially become the first woman to make the crossing by air, but she had been a passenger and not the pilot. It was not until 1932 that she made her historic solo flight. Ruth Elder's abortive attempt had taken place in 1927, while Ruth Nichols tried and failed to make the transatlantic crossing in 1931.

The Air Derby was far more than a mere competitive event, as its first winner, the twenty-three-year-old Louise Thaden, realised, although her description of it was rather over the top: 'The successful completion of the Derby was of more import than life or death.' In a sense it proved that flying – for women just as for men – could be safe, because so many competitors completed the difficult course. However, one entrant, Marvel Crosson, was killed, baling out too late and too low for her parachute to open properly, and critics of female flyers seized on this well-publicised tragedy to underline their point: 'Women Have Conclusively Proven That They Cannot Fly', as one newspaper put it.

An indirect result of the Derby was the formation of an organisation of licensed women pilots. Inspired by the sense of shared dedication and – in Amelia Earhart's words – the 'concerted activity' which it underlined, preliminary meetings were called in November and December 1929: of the 126 licensed female pilots ninety-nine decided to join the new association, which appropriately adopted the name of The Ninety-Nines, and almost immediately became a powerful influence in aviation. Amelia Earhart was its first president and it has gone from strength to strength over the decades as an educational and lobbying force.

There is no doubt that America became the special focus of aeronautical achievement towards the end of the 1920s, partly of course, as a result of Charles Lindbergh's May 1927 crossing of the North Atlantic. Britain's John Alcock and Arthur Whitten-Brown had achieved the transatlantic flight as early as in 1919, but their route was shorter than Lindbergh's Long Island to Paris one, and he was the first to make it solo. His lonely battle with the hazards of the trip in the small Ryan monoplane *Spirit of St Louis* not only won him the $25,000 Orteig Prize – which had been on offer for the New York to France crossing since 1919 – but made him the hero of millions, both in America and all over the world. Despite his good looks, confidence and general air of distinction, Lindbergh was an uneasy figure when the international limelight turned upon him. Before making his epic flight at twenty-five years of age, he had earned a living first through stunting and barnstorming, then as a mail pilot. His interest in widening the frontiers of aviation was intense, but he hated much of the publicity which was foisted on him in 1927 and for much of his subsequent life. Puritanical, frequently controversial and soon established as America's – if not the world's – most eligible bachelor, he married in 1929 the small, shy and serious-minded Anne Morrow, the daughter of Dwight Morrow, an influential and highly successful banker, lawyer and diplomat.

Charles quickly coopted Anne as a crew member on some of his flights, and she became a skilled and enthusiastic airwoman, although her tastes were primarily literary and studious. The Lindberghs' money and fame provided them with many opportunities for effecting aviational innovations and improvements, and together they set new records and opened up new routes. Anne became the first woman in America to obtain a glider pilot's licence and she also qualified as a radio operator. She relished the exhilarating sense of freedom that she and Charles always felt when they were in the air and away at least temporarily from the glare of media attention. She was also at pains to demonstrate how safe flying could be, and, possibly with this particularly in mind, continued to fly throughout most of her first pregnancy.

The tragedy of the kidnap and death of their first baby, Charles junior, and the resultant overwhelming and long-drawn-out publicity, underlined the Lindberghs' heroic reputation. Their popularity survived their departure from the USA to take up residence in England from the end of 1934 – although it became slightly dented when Lindbergh's frequent trips to Germany between 1936 and 1939 as a guest of the Nazi administration made him an advocate for some of its policies and an ardent believer in the advisability of maintaining a strong Germany as a buffer against communism.

Returning to America after the outbreak of the Second World War, Lindbergh offered to serve in the Army Air Force, but was rejected. He then worked with Henry Ford to increase aircraft production, became an adviser to the USAAF in the the Pacific area and, at the end of the war, revisited Germany to try to recruit their most prominent aviation and rocket experts for the West. He and Anne had always hoped that aviation would bring the nations of the world together, but by the 1960s Charles was becoming more and more disillusioned about this. The Lindberghs devoted much of their later life to conservation and support of the World Wildlife Fund. Charles died at the age of seventy-two in 1974 at the home they had built on the Hawaiian island of Maui, which had for some time been their haven from disenchantment and unwelcome publicity.

Anne Morrow Lindbergh established a reputation as a writer, and her poems, particularly in *Gift from the Sea*, provided glimpses of the rich inner life which sustained her as both flyer and woman during the vicissitudes of her life with Charles. Simplicity and the need for solitude are keynotes, as well as emphasis on space, both psychological and physical: 'Only in space are events and objects and people unique and significant – and therefore beautiful. Even small and casual things take on significance if they are washed in space.'

When Amelia Earhart became the first woman to be flown across the Atlantic in 1928, Charles Lindbergh's heroic image was at its most intense in the public eye. Amelia was to become a legend in her own right, but initially the process was helped and speeded by her marked physical resemblance to Lindbergh. Both were tall, slim, fresh-complexioned and blue-eyed, somewhat reserved, but with occasional wide, flashing grins. George Palmer Putnam, the publisher who promoted Amelia's first international flight and was later to become her husband, capitalised on the likeness for publicity purposes. She was dubbed 'Lady Lindy', and the nickname, used extensively by Putnam and the media, stuck. Subsequently, of course, Putnam was accused of manufacturing the resemblance, but Anne Lindbergh's considered comments in her book *Hour of Gold, Hour of Lead* (1973) seem to give the lie to this: writing of Amelia, she says, 'She is the most amazing person – just as tremendous as C, I think. It startles me how much alike they are.'

Amelia Mary Earhart was born on 24 July 1897 in Atchison, Kansas. Her father, Edwin Stanton Earhart, was a young lawyer whose career was to become blighted to an extent by his hard drinking. Her mother Amy, the daughter of a prominent judge, was an early feminist who brought up Amelia ('Millie') and her younger sister Muriel ('Pidge') with a great deal of freedom and many interests beyond the then conventional domestic ones. After a visit to the 1904 World Fair in St Louis, Amelia's early interest in mechanics prompted her to design and build a roller coaster in the back garden of her home, using long planks propped against the roof of the tool shed. Amy was sensible enough to encourage her daughters in their outdoor and tomboyish games. At a time when small girls wore full-skirted dresses and frilly pinafores, she even had bloomered gym suits made for them. Both girls rode, and Amelia became an excellent horsewoman.

As a child Amelia read widely. As well as classics by Alexandre Dumas, Victor Hugo and others, she enjoyed boys' adventure stories, which she savoured in periodical magazines, animal stories and a book called *Insect Life*. She was a good scholar, though not of a particularly studious disposition. The family life of the Earharts became unsettled, with temporary separations between Amy and Edwin and, eventually, a divorce. The girls' schooling was interrupted by several moves, and Amelia's plan to go to college was interrupted by the events of the First World War. A visit to Toronto for the 1917 Christmas holidays brought her into contact with many of the severely wounded soldiers who were being nursed there at the Spadina Military Convalescent Hospital. Much moved by their plight, Amelia joined a Voluntary Aid Detachment and worked at the Spadina Hospital until the November 1918 Armistice.

She fell victim to the terrible influenza epidemic which swept much of the world towards the end of the war, and was further weakened by a severe sinus infection, which gave her trouble, on and off, for many years. Her hospital experiences had convinced her that she would like to become a doctor, and she enrolled as a medical student at New York's Columbia University. She opted out of this fairly soon, however, partly because she felt unsure about medicine as a career after all, and partly because Amy was putting pressure on her to return to the family home, which was now in Los Angeles. Her parents were undergoing one of their reconciliations (before what turned out to be the final divorce) and her mother felt that Amelia's presence might help to bind their marriage.

At twenty-three Amelia was tall and slim, with fair, very long hair which she was soon to cut short, inch by inch, so as not to alarm her family too much. She and Sam Chapman, a young chemical engineer who was boarding in the Earhart home, became attracted to each other. He stirred her social conscience into vigorous action and they attended various socialist meetings together. However, campaigning for social and political reform took a back seat after Amelia attended a Long Beach air show in 1920, was taken up for her first $10 flight and decided that she would have to find at least $500 for a course of flying lessons which she began under the tutelage of a woman instructor, Neta Snook. With unusual impulsiveness, Amelia decided, after only two or three hours' tuition, to buy her own aircraft, a Kinner sportsplane which would cost $2000. She received some financial help from her family, but had to take twenty-eight different jobs in two years to raise the money which she still required for her aviational activities. These ranged from clerical work to driving a heavy truck for a sand and gravel company: Amelia was nothing if not determined. She spent most of her free time at the Long Beach airfield, and adopted a style of flying dress which greatly suited her: it was basically similar to that of most male pilots, consisting of boots, breeches and a long leather flying-jacket (the shiny newness of this stimulated some teasing from workers and spectators at the airfield, so she prematurely aged the coat by deliberately staining and sometimes sleeping in it).

Amelia's first solo flight took place in 1922 and, soon afterwards, she flew to a height of 14,000 feet to create a new world altitude record for women. Only a few weeks later Ruth Nichols went even higher, and in trying to re-establish her record Amelia became engulfed in dense fog at 12,000 feet; she nearly lost her life by coming out of this in a deliberate spin – which was, she said, 'the quickest way down' known to her. She was criticised for her recklessness in this, for the clouds might well have closed in on her right to the ground, and also for the slight tendency to daydream during flight which Neta Snook observed during her early lessons.

She found flying not only satisfying in itself but a useful escape from her parents' marital problems and from the pressure which Sam Chapman was putting on her after proposing marriage. Although they were supposed to have been engaged for a period, she was extremely unsure about embarking upon the fairly conventional marriage which he envisaged.

Despite its fulfilments, however, it seemed that flying could not provide Amelia with a living. She continued, of course, to go up for pleasure, but commercial aviation jobs were beyond her reach. Social work seemed to offer an attractive means of earning a living: 1927 found her helping to solve the emotional and educational problems of immigrant children, in a $60-a-month job at Boston's Denison House.

The tenor of Amelia's life changed dramatically when she received a surprise telephone call in April 1928 asking whether she would consider taking part in a transatlantic flight as a passenger. The expedition was to be financed by Mrs Amy Guest, the affluent American wife of Captain F. E. Guest, the Liberal MP, who from 1921 to 1922 had been Britain's Secretary of State for Air. Mrs Guest had purchased her three-engined Fokker F.VIIB/3m *Friendship* in the hope of making the flight across the Atlantic herself, but her family had dissuaded her from this dangerous undertaking. George Putnam, who was to become the flight's manager, entrepreneur and publicist, had been charged with finding a suitable young American woman to take Mrs Guest's place, and Amelia's name had been suggested by a retired rear-admiral from Boston, Reginald K. Belknap. The lady in question had to be sufficiently cultured to be able to hob-nob with British aristocrats and dignitaries after the hoped-for successful landing, but also of a practical and robust nature, with, of course, some understanding of aviation. Amelia, who had been described as 'a young social worker who flies', seemed perfect for the job.

She knew from the outset that her role was to be an inactive one. The piloting would be done by the hard-drinking veteran Wilmer ('Bull') Stultz, while Louis ('Slim') Gordon would navigate. She was also only too well aware of the dangers. In the previous year alone, which was marked by Lindbergh's solo crossing, nineteen people had died in their attempts. Two of these had been female passengers, each hoping, like Amelia, to become the first woman to cross the Atlantic by air: the planes which carried the American Frances Grayson and the British-born Princess Anne Loewenstein-Wertheim both vanished without trace not long after taking off.

At Trepassey Bay, their point of departure, Amelia and the two pilots had to endure a nervous two-week wait for the appropriate weather conditions. Amelia's natural anxieties about the flight turned to alarm when Stultz had eventually to be lugged into the cabin, fuddled with drink, for take-off on 17 June. Nevertheless, despite intense cold and drifting fog, and lack of space and sleep, the crew and passenger of *Friendship* landed safely at Burry Port in Wales (expecting it to be Ireland) after being in the air for twenty hours and forty minutes.

The wild and wonderful welcome which they received was focused strongly on Amelia, in spite of her efforts to stress that all the credit for the flight should be reserved for Stultz and Gordon. She said that she had been 'as useful as a sack of potatoes'. However, she was considerably more charismatic – and helpful for publicity purposes – than any other kind of cargo. The British public's rapturous response to Amelia was conveyed at great civic receptions, by cheering crowds and invitations to attend a 'jumble of teas, theatres' and sporting events. She was allowed to shop at Selfridges in the West End of London free of charge, and she visited Toynbee Hall, in the East End, because Denison House in Boston, where she had been working, was modelled on this University of London settlement. She was received by several aristocratic families, and, in popular near-fairy-tale tradition, danced with the Prince of Wales.

The press and radio reporters generally went crazy over Amelia, but, inevitably, there were one or two waspish comments. The editor of the prestigious British aviation magazine *Flight* wrote wryly that the first lady passenger's crossing of the Atlantic by air meant little by comparison with the recent solo flights by 'such lady *pilots* as Lady Bailey and Lady Heath'. A *Church Times* article pointed out that Amelia had become a heroine 'solely because, owing to good luck and an airman's skill and efficiency, she is the first woman to travel from America to Europe by air'.

In fairness to Amelia, she was really saying exactly the same thing. And of course it *did* require courage to risk one's life in a flimsy aircraft over such a long stretch of ocean. Her fears and bravery are underlined in letters to her parents which she wrote before take-off and entrusted to Putnam for delivery in the event of her death. They were short and very much to the point. The note to her father said, 'Hooray for the last grand adventure! I wish I had won but it was worth while anyway. You know that.' To her mother she wrote, 'Even though I have lost, the adventure was worthwhile. Our family tends to be too secure. My life has really been very happy, and I don't mind contemplating its end in the midst of it.'

As a final British response to Amelia's Atlantic crossing, an article in the 10 November 1928 issue of the paper *School-Days* should be mentioned. Under the heading 'Schoolgirls are Proud of' she gets only a few words. British girls, it seems, should appreciate the American aviatrix 'for being the first woman to fly across the Atlantic Ocean and for going home quietly afterwards to settle down to welfare work'.

Of course, there was nothing quiet or settled about Amelia's homecoming in America, or in her future. 'Lady Lindy', under Putnam's management (though she would have done very well without it anyway) became a heroine, 'a symbol of new womanhood' and an up-market role model for those whose American dream was of becoming career women. She attended dinners, lectured, became aviation editor of *Cosmopolitan* magazine, and spoke out robustly for feminism and for greater opportunities for girls in aviation.

Her somewhat tenuous engagement to Sam Chapman was finally ended, and after George Putnam was divorced in 1930 he submitted her to a barrage of proposals. She finally accepted him, and they were married in February 1931. Amelia had doubts about giving up her single, independent state, despite their mutual attraction, and presented Putnam on their wedding day with a letter setting out her terms for their partnership. It was honest, dignified, liberal and unexploitive: 'I shall not hold you to any medieval code of faithfulness to me, nor shall I consider myself bound to you similarly. . . . Please let us not interfere with the other's work or play, nor let the world see our private joys or disagreements.' She also asked George to promise to release her in a year if they should not find happiness together.

Their marriage survived, and Amelia's flying achievements in the six years which were left to her were truly outstanding. In 1932 she confounded the critics who had dismissed her 1928 flight across the Atlantic by becoming the first woman to cross it flying solo. (In fact no one apart from Lindbergh had yet made the flight alone.) As well as becoming the first president of The Ninety-Nines in 1930, she won the Harmon Trophy as America's Outstanding Airwoman for three consecutive years; in 1935 she became the first person to fly solo across the Pacific Ocean from Honolulu, Hawaii, to Oakland, California (2408 miles) and the first person to fly solo non-stop from Mexico City to Newark, New Jersey. Congress awarded her the Distinguished Flying Cross; the French government conferred upon her the Cross of Chevalier of the Légion d'Honneur and President Hoover presented her with the Gold Medal of the National Geographic Society.

In 1937 Amelia decided to tackle a round-the-world flight at the Equator, which no one had previously attempted. Purdue University, where from time to time she had been the 'resident aviatrix' and careers' consultant for girl students, made available $50,000 for a 'flying laboratory'. She arranged to fly in a Lockheed Electra, an advanced twin-engined aircraft with dual controls. This plane was often used in civil aviation; it could carry ten passengers as well as two pilots. The passenger seats were ripped out so that a great variety of instruments and devices could be fitted to the Electra, which could fly at 210 miles per hour, with a top altitude of 27,000 feet. Amelia's aerial laboratory was lovingly described by the press; it included blind-flying instruments, a Sperry auto-pilot and complex radio equipment which should ensure the safety of Amelia and her only companion on the flight, Fred Noonan, her navigator.

Noonan was another of American aviation's colourful characters. A navigator who had survived being torpedoed in the First World War, he had helped Pan American Airlines in its early days to set up its Pacific routes. He was said to have made eighteen crossings directing the flight of the company's *China Clipper*: certainly his knowledge of the Pacific was extensive.

If Amelia's flight succeeded, she would be the first woman

to complete the round-the-world trip. Wiley Post had ringed the world in 1931 in a Lockheed Vega, with Harold Gatty navigating, and followed up this achievement by doing it solo in 1933. He had unfortunately been killed in a crash in 1935, in company with the celebrated American comedian, Will Rogers. (One of Post's contributions to aviation was the pioneering of an early pressure-suit for high-altitude flights.)

Amelia was not setting out to break any speed records on this trip, but wanted to travel the longest possible distance by traversing the world around the Equator. Apart from the satisfaction of completing the arduous journey, she claimed that her main motivation was to make 'a thorough check of modern equipment'. She would keep a detailed log of what happened to 'the personnel' (herself and Noonan) and the machine under various conditions: 'Records such as these, be they of success or failure, can do much to safeguard subsequent flights.'

The 'flying laboratory' was officially presented to Amelia on her thirty-ninth birthday. From July 1936 until March 1937 she worked diligently to familiarise herself with all the complexities of the plane's instruments and, particularly, the blind-flying trainer. Putnam did his usual thorough preflight publicity job, and Amelia took off in the early hours of 17 March. Unfortunately on the second leg of the journey the Electra crashed on the runway; it had to be shipped back for extensive and costly repairs, but Amelia firmly announced that she would try again. Finding all the necessary funds for the trip had been difficult enough even without the crash; now at least another $50,000 would be required. Happily help was forthcoming from various friends and admirers, including the aviatrix Jacqueline Cochran and her millionaire husband Floyd Odlum. The two aviatrices had become friends, a special link between them being their mutual interest in psychic and extra-sensory matters. On one or two occasions Amelia had claimed that her own extrasensory powers had helped in the finding of wreckage after plane crashes. She and Jackie Cochran had made a pact to contact each other psychically should either of them need help in an emergency.

By the time the Lockheed Electra was ready for its second take-off – from Miami on 1 June – changes in the weather conditions necessitated alterations in the route; it was decided that Amelia's navigator, Fred Noonan, would now be with her all the time, although on the first schedule it was planned that the aviatrix would fly some laps alone. All went reasonably well on the first 22,000 miles of the flight, during which they crossed the Equator three times and landed on twenty-two occasions, taking such hazards as monsoon storms over India confidently in their skyward stride.

On 2 July they took off from Lae, New Guinea, for Howland Island (where they would have to refuel), knowing that this would be the most difficult part of their journey. Even the slightest error could mean their missing Howland, a flat Pacific island which was only 2 miles long and half a mile wide. As back-up, the US navy tug *Ontario* was positioned half-way along the route, the coastguard cutter *Itasca* stood by just off the island, and the navy ship *Swan* waited between Howland and Hawaii.

The *Itasca* was equipped with a radio direction-finder and a radio beacon that could be picked up by the Lockheed's Bendix Radio Compass. Once the plane came within a couple of hundred miles of Howland, the ship could guide them in. That, at any rate, was the plan, but things went disastrously awry; the radio connection broke down when the Lockheed was some 100 miles off course and in imperative need of bearing correction. Regular and then spasmodic radio messages were being received by the *Itasca*, but it became obvious that Amelia was not receiving the responses that the ship was sending. Wave-lengths were changed, homing signals were broadcast in frantic efforts to help Amelia and Noonan to get back on course, but all was to no avail. Conditions were cloudy and overcast; static – suggestive of a storm brewing – was beginning to obscure the radio messages coming from the plane. The last to be heard clearly was at 8.44 am, local time: 'We are on the line of position 157–337, will repeat this message on 6,210 KCS. . . . We are running north and south.'

There was no repeat of this, and subsequent supposed

messages from the flyer and navigator, suggesting that they were floating in the Pacific on the debris of the Lockheed, were established as hoax calls.

President Roosevelt set in motion a search of the area. It lasted two weeks, scoured 25,000 miles of the Pacific and cost approximately $4 millon. No trace of nor clue to the where-abouts of the Lockheed Electra and its crew was found. Of course, there were criticisms of the way in which the search was conducted – that it started too late, was not properly organised, did not cover the correct ground, etc. George Putnam was over-ready with advice as to which areas should be covered; some of his suggestions derived from Jackie Cochran whose extra-sensory capabilities apparently indi-cated that the plane remained afloat and Amelia and Noonan were alive, although he had sustained severe head injuries. Jackie gave the name of the *Itasca* as well as that of a Japanese fishing boat which, it transpired, *was* in the area – but after two days she felt sure that Amelia and Noonan were dead.

There were suggestions that the US government was as much concerned with finding out about Japanese fleet pos-itions and activity as with locating Amelia. Putnam and others hinted that the long search for the lost pilot and navigator was an excuse for US ships and planes to make a thorough exploration of the area. This was perhaps the first of the many rumours which began after Amelia's disappear-ance: it soon became obvious that she was going to be far more than just a legend in her own life-time. A whole industry of 'What happened to Amelia Earhart?' sprang up, and became self-sustaining across five decades. Exploration of the many theories would make a book in itself (indeed it has already made several). One of the most bizarre sugges-tions is that Amelia was on some top-security mission for the US government, after which she wanted to disappear from public life; that she was still alive, and resident in America under an assumed identity (and not, apparently, wanting to fly any more). Another recurring-with-many-variations theory was that she was on an espionage assignment for the US government, and that she and Noonan had been captured by the Japanese and shot. After certain supposed sightings of the American pair, reported by local inhabitants, exhumations of bones were made years afterwards. The only thing these proved was that the bones had nothing to do with Amelia and Fred.

The rumour about Amelia being on a spying mission has persisted over the years, although in view of her outspoken pacifism, it seems unlikely that the US government would have cast her in such a role. Nevertheless, this is the image which seems most firmly fixed in the public's assessment of the affair. Credence appeared to be given to it when RKO released a film called *Flight for Freedom* in 1942, starring Rosalind Russell as a 1930s' aviatrix and Fred MacMurray as her navigator. The heroine, Tonni Carter, who was obviously based loosely on Amelia, was asked by the US government to use her round-the-world trip for under-cover activities. The plot suggested that she should land secretly on a small Pacific island and pretend to be lost – thus enabling the Americans, in their supposed search for her, to fly over Japanese-mandated territory to inspect fortifications, airfields etc. The film was made at a time when America and Japan were at war, and the public would have viewed Amelia's involvement in this kind of activity as an underlining of her heroic image. Nevertheless, it seems unlikely that the plot of *Flight for Freedom* provided a true echo of Amelia's last journey.

George Putnam was sufficiently sure that his wife had died soon after the plane disappeared to marry again eighteen months afterwards. In 1991, the long-drawn out speculation about Amelia's last days seemed to be ending, as some hard evidence at last emerged. An FBI statement was made that a suitcase-sized aluminium box had been recovered from Nikumaroro (formerly known as Gardner Island). The aviation part number stamped in the box positively identified it as a bookcase for stowing navigation aids, and as part of the equipment of Amelia's Lockheed Electra. Nikumaroro is about 60 miles from Howland Island; tragically it has no fresh water, so if Amelia and Noonan were grounded on it they would have died of thirst.

Much still has to be uncovered before the speculations can be laid to rest, but apparently the division of Oceaneering International, which recovered the debris of the space shuttle *Challenger* after it exploded in January 1986, will be searching for further evidence as soon as the necessary amount of money (something like £200,000) is raised. Whatever is or is not discovered, of one thing we can be certain: the legendary Amelia Earhart will never really die.

Her critics have questioned her ability as a pilot and have accused her – and Putnam – of arrogance. There is, however, no doubt that her contribution to aviation during the inter-war years was unique, both symbolically and in fact. Despite the legends and the ballyhoo which often surrounded her, she was basically unpretentious and a realist; the way forward, as she saw it, was simple: 'Women must try to do things as men have tried, and their failure must be but a challenge to others.'

It was probably Amelia's popularity which triggered the creation of 'Flyin' Jenny', the first American comic-strip solo aviatrix, in 1939. Just like Amelia's, Jenny Dare's flying clothes (jodhpurs, long boots and so on) were reminiscent of the early male military flyers. Also like Amelia, Jenny was to become something of a legend. Her name was derived from the Curtiss JN-4 training plane of the First World War, and the links with the early days of aviation provided a pleasant touch for the strip, even though Jenny's exploits were supposed to have taken place from the late 1930s until the mid-1940s.

The experienced cartoonist Russell Keaton, who had worked for some time on the popular 'Skyroads' strip, proposed the idea of an Earhart-type heroine to the Bell Syndicate, who approved and quickly got her off the drawing-board and into production. At first Keaton produced both the pictures and the storylines, but later he entrusted the text to Frank Wead, a former navy pilot and a Hollywood scriptwriter. War service claimed Wead in 1941, so Glen Chaffin, who had scripted 'Tailspin Tommy', another popular flying feature, took on the writing of the stories.

For some of the war years, Keaton worked as a flying instructor, and one of his assistants, Marc Swayze, did much of the drawing. He took over completely when, in early 1945, Russell Keaton was diagnosed as having acute leukaemia and died soon afterwards (he was only thirty-five). Flyin' Jenny's exploits ended in October 1946, fairly soon after Keaton's death. It seemed that although there was a place for the blonde-daredevil pilot in the 1930s and early 1940s, she was, like so many real-life women and girls in the immediate post-war years, expected to get out of her aeroplane and back to the kitchen sink – 'Sorry, Jenny, pilots are a dime a dozen' – leaving the serious jobs in civil aviation for the boys once again.

Jenny starts off as every inch the confident career girl, quickly winning the respect and admiration of her male peers. She is drawn as a vibrant and sexually attractive young woman, but firmly slaps down any male who is rash enough to express appreciation of her charms. (In this respect she responds as robustly as W. E. Johns's Worrals, whom she pre-dates by a year or two: see chapter 9.) She flies all kinds of aircraft in all kinds of weather, is sometimes a test pilot and sometimes an air detective. In between career assignments, she finds time to take place in cliff- (or sky-) hanging races and competitions.

One of Jenny's greatest aviational skills is the making of difficult landings: in her first strip, when she is after a test-piloting job with Starcraft Aviation Factory, she shows off a bit by 'coming in for a landing – *upside down*.' By the 1940s, when she is searching for Nazi looted treasure on a remote Pacific island, she has lost none of her panache or technique, as a male colleague points out: 'Nobody but Jenny could pillow-land that awkward old crate like that. She's okay!'

Her pre-war adventures sound staid on the surface (investigating fraudulent insurance claims, for example), but they are actually lively amalgams of sabotage, shipwreck, murder and bitchy rivalry between female flyers (a distinctly un-Earhart touch!). She looks fearfully fetching with curls peeping out from her flying helmet and curves softening the vintage macho flying garb. As the 1940s progressed, she tended to swap her authentic-looking outfits for flesh-revealing cut-away ensembles, such as bare-midriffed two-pieces. She stuck, however, to her 1930s' jargon, ejaculating 'Jeepers!' in moments of stress or exhilaration and labelling everyone whom she likes as 'cute'.

Her contribution to the war effort was the organising of a 'bird-girl shuttle command' to ferry important cargo. The last two or three stories lack some of the exuberance which was so integral an ingredient of the early strips. Her difficulties – and those of some colleagues and friends – in adjusting to the post-war world ('H-mm! War in China . . . Europe a mess . . . Reconversion problems scaring Americans and British') were a signal that fresh challenges would lie ahead for aspiring and determined aviatrices.

CHAPTER SIX

Amy, Wonderful Amy!

FREE!

Beautiful
Stand-up
Photo of

NOW ON SALE

AMY JOHNSON

GIVEN AWAY
WITH THIS WEEK'S

SCHOOL·DAYS

NOW ON SALE

The Chummy Paper for Schoolgirls.   Price 2ᴰ

There is nothing more wonderful and thrilling that I can imagine than going up into the spaciousness of the skies in a tiny light plane where you feel alone, at peace with everyone, and exactly free to do what you want and go where you will.

AMY JOHNSON, broadcasting after her
1930 England to Australia flight

B Y THE beginning of the 1930s airmindedness was becoming a regular theme in newspaper and magazine articles in Britain as well as in America. Even before Amy Johnson made her celebrated solo flight from Britain to Australia in May 1930, readers of both sexes were being encouraged to feel that the exciting, expansive world of aviation would be accessible to everyone and not the exclusive preserve of a privileged few. (In fact, of course, cheap air travel was not to come for another two or three decades, because developments in civil flying were to be drastically curbed during and just after the period of the Second World War.)

In an attempt to echo the popular mood, the rather conservative and up-market shilling monthly *Woman's Magazine* ran an article in April 1930 called 'Keeping an Aeroplane of One's Own'. This was written by Sicele O'Brien, a plucky pilot who continued to fly even after losing a leg in a crash. She suggested that 'a great many people would learn to fly, and keep their own private aeroplanes' if they were not scared off by 'exaggerated ideas of the cost, and of the imaginary difficulty of housing such a large and beautiful object with its wide-spreading wings'. She blithely explains that most modern light aeroplanes 'can be folded or opened by a girl single-handed in less than two minutes', and that, when folded, the machine can be accommodated in an average-sized garage. It is, apparently, a comparatively easy matter to establish a 'private aerodrome on one's own estate', but should there be too many trees, telegraph wires or houses in the way of take-offs and landings, one of the 'flying clubs springing up all over the country' would be happy to offer its facilities. Sicele O'Brien goes on to explain that modern machines are fairly easy to maintain 'if one has even a wee bit of a mechanical mind'. It is simply a question of emptying grit or water from the carburettor filter, cleaning the sparking plugs, adjusting the tappets and oiling the valve rockers.

The article is headed with a photograph of a well-turned-out, cloche-hatted and purposeful young woman pulling a dapper folded-wing Moth from its garage behind her. This looks easy enough to do with just one hand – but one suspects many flaws in Sicele O'Brien's persuasive text. For a start, of course, in 1930 very few families could even afford cars, let alone aeroplanes, and without a car it might well be very difficult to make use of all those flying club facilities which she mentions. Although the idea of Britain liberally bespattered with such clubs is attractive, in fact they were then few and extremely far between, as well as being beyond the price range of most of the readers of *Woman's Magazine*.

For a younger female audience, 1930 editions of *School-Days* provided further aviational snippets, with a photograph of Eleanor Smith, a seventeen-year-old American 'who flew her plane to a record height', and a picture of three young aviatrices from the Far East: 'The girls of Japan are no longer those rather wistful, quiet people'; now, 'like the girls of other nations they are always doing daring stunts and winning races in the air'.

Fetching photos of women flyers were increasingly used to promote the sales of everything from toothpaste to tarpaulin, while magazines – which often had nothing to do with aviation – pulled in readers through engaging cover pictures of girls in flying garb standing beside inviting-looking planes. (Arthur Ferrier's 'Plane Jane' on the cover of *Ideas* (27 January 1934) was one of the most striking of these.)

Flying stories for men and boys continued to flow from publishers throughout the country, but there was still a certain ambivalence about similar stories for women and girls. Although flying was seen as a general symbol of liberation, editors were reluctant to link aviation with serious female ambitions. It was acceptable to do so in fiction aimed at a lower age level, presumably because young girls were allowed a measure of fantasy in their reading. Nevertheless it was expected that once a woman had reached maturity, she would prefer to read stories which reflected down-to-earth realities – the finding of a husband, the establishment of a home, the triumphs and tribulations of motherhood. There was very little fiction which focused seriously on careers, because most women gave these up on marriage, anyway. Even at the escapist level, aviation didn't get much of a showing: rightly or wrongly it was felt that this kind of fantasy was better provided in high-flown romances and historical sagas.

Sometimes aviatrices got a look-in as characters in male-oriented stories. For example, John Bolton's *The Mystery Plane* (a title in the Mellifont Press's threepenny Air Stories Series) featured a flyer called Peggy Lee Fox as well as the main RAF hero, Jack Hubbard. The daughter of his CO, Peggy is, according to Jack's batman, Alf, 'pretty as punch, and with five hundred officers yearning their blooming eyes after her, all she wants to do is fly'. She is, apparently 'a masterly pilot' and 'If she had not yet crossed the Atlantic or made a record flight to Australia, it was no fault of her ambition.' She spends 'all her spare time flying, studying aircraft design, watching aeroplane engines being dismantled or reassembled'. She is a curious mixture of independence and fluttering subservience to the male. Having persuaded her outwardly macho but soft-centred father, Sir Evelyn Lee Fox, to buy her an aeroplane (of an unspecified design) she wheedles from him 'a new sports car'. At the same time, she plans to make her own way in the world and resents any attempt from Jack (who fancies her when he sees her out of flying gear and in skirts) to foist upon her the little woman's traditional role. Whenever she has the opportunity, she delights in shocking him 'by her calm acceptance and talk of things which he thought nice women never mentioned, even if they knew of their existence'. Most shocking of all, perhaps, is the fact that when Jack finds himself in an RAF display up against a particularly fiendish opponent in a mock air battle, 'he' turns out to be Peggy: 'Some pilot, that,' his observer murmurs, as they eventually climb shakily out of the cockpit.

Jack Hubbard is on an undercover mission at the RAF station, sniffing out the aerial activities of a particularly unsavoury Teutonic spy suspect called Imhof. It is made clear that this character is *not* in the pay of the German government (the story was written in the days when we were still trying to appease the Nazi dictator, who was then always respectfully referred to as Herr Hitler). He is simply motivated by 'a hatred of the English, allied to a megalomania which made him picture himself as a kind of World Dictator, with England and the British Empire ground beneath his heel'. Unfortunately, at one point in the story, Peggy finds herself in his power. Luckily Hubbard realises just in time that Peggy has had to make a forced landing at Imhof's private aerodrome and races there to rescue her:

> The big hangar, lit by arc lamps, presented a spectacle to make any man's blood run cold. Imhof was turning the airscrew of a light aeroplane. Tied from blade to blade of the propeller, like a victim at the stake, was Peggy. At any instant, the engine might roar into life!
>
> With a terrible yell, Hubbard hurled himself at the monster . . .

Not unnaturally, Peggy is less cheerfully voluble than usual after her ordeal, and Hubbard, who had expected hysterics, asks if anything is troubling her. 'Very shyly', she whispers, 'I've got a funny feeling . . . I want to stroke your hand.' On the tide of this passionate declaration they decide to marry, with Peggy promising not to invite danger by taking to the skies again until at least 'after the honeymoon'.

Amy Johnson achieved the first solo flight by a woman from England to Australia. With surprisingly few flying hours behind her, she left Croydon Aerodrome on 5 May 1930 and arrived at Darwin on the 14th. Her determination, endurance and general charisma caught the imagination of the public not only in Britain and Australia but all over the world. Although appearing at first to enjoy the adulation resulting from her success, she was to find it increasingly hard, as her career progressed, to cope with such intensities of feeling. She was disturbingly conscious that the frenetic praise which she attracted sprang less from her technical achievements than from surprise that a female could cope with the challenges of long-distance flights. In spite of receiving tributes from other aviators, from statesmen, movie stars and crowned heads, the 'Amy, Wonderful Amy' of popular song found it almost impossible to obtain a serious flying job.

The daughter of a comfortably-off Hull fish merchant, she was born in 1903 and brought up with her two sisters Irene and Molly according to Methodist principles. At fourteen she suffered the trauma of losing several front teeth after being hit in the face by a cricket ball, and there is no doubt that, despite the fitting of a good denture, she felt disfigured and unhappily self-conscious for some years afterwards. She was to blame this accident for the moodiness and over-sensitivity which dogged much of her life.

There certainly seemed always to be something extremely vulnerable about Amy. Just before becoming a student at Sheffield University she had fallen in love with a Swiss (whose true name was never revealed by Amy or her biographers, but who is always referred to as Franz). Eight years older than Amy, he seems never really to have been serious about her, but she worked desperately hard on the relationship, convinced for several years that he would marry her. He did not – because he found someone else whom he preferred, who seemed more likely to fit in with his Catholic upbringing and whom he fairly quickly made his wife.

It is not exactly true to say that Amy took up flying on the rebound from her unhappy affair with Franz, but it seems likely that her determination to achieve success in a very definite career rather than in marriage became sharpened by the frustrations in which the relationship involved her.

Getting an interesting and well-paid job after obtaining her degree was no easy matter, and she was forced to take a series of fairly low-level and unsatisfying positions from shorthand-typing in a local accountant's office to working at Peter Jones department store in London. She had a spell with an advertising company and in a solicitor's office where she became secretary to a senior partner, showing flair for the work. It seems significant, however, that in all these jobs Amy felt isolated from her work-mates and appeared by turns arrogant or inadequate.

When Amy first experimented with flying in 1928 she still had tenuous hopes of Franz: when these were finally dashed by his marriage, she toyed with the idea of suicide by apparent misadventure in the air, but fortunately failed to follow this up. She had to wait some time for her first flying lesson at the London Aeroplane Club in Stag Lane, Hendon. There was a long waiting-list for these government-subsidised lessons which, at £1 10s an hour, Amy could just about afford from the £5 a week (no mean wage in 1928) which she earned in the solicitor's office. Her first period of instruction, from Captain F. R. Matthews in a DH 60 Cirrus II Moth, was distinctly disappointing, and he told her uncompromisingly that she was no good. The techniques that Lady Heath had once declared to be 'ten times easier than riding a bicycle' seemed to elude Amy. Happily she achieved greater rapport with the Club's chief flying instructor than with Matthews. He was Captain Valentine Henry Baker, MC, AFC (who was to be killed when test-piloting during the Second World War). Baker gave Amy five out of her ten initial lessons and, with a reputation for teaching the unteachable, he inspired her with confidence that helped her to acquire both her private and commercial pilot's licences.

At that time, Lady Heath was the only British woman to have formally qualified as a ground engineer, but she had done so in the United States. Jack Humphreys, the chief engineer at Stag Lane, spotted Amy's flair for mechanics and gave her a great deal of help and encouragement. Thus Amy achieved the distinction of becoming the first female ground engineer to qualify in Britain. (It is no wonder that she sometimes became irritated when, with this, her degree and her hard-won pilot's licences behind her, people still used to refer to her as the typist who became an airwoman.)

With the help of Jack Humphreys, Amy was able to persuade her father, Will Johnson, to give her sufficient financial support to enable her to devote all her time and energies to flying. Her ambitions soared, and early in 1930 she began to plan her solo flight to Australia. Finding sponsorship seemed a marathon task, but eventually her father agreed to help with the cost of a suitable plane, and Lord Wakefield, the oil magnate who gave generous help in many areas of aviation, provided the rest.

Amy acquired a two-year-old DH 60G Gipsy Moth from Captain Wally Hope, an experienced flyer. The machine was already fitted with special long-distance fuel tanks and had proved its reliability on one journey to the tropics. She proudly named it *Jason*, which was a contraction of 'Johnson', the registered trademark of the family's Hull-based fish business. She had it painted bottle-green (her then favourite colour), with its name picked out in silver.

Captain Hope appears to have given her a great deal of advice, not only about handling the machine but about the equipment which might be advisable for the trip. First in priority was a revolver, then a mosquito-net and a sun-helmet, plus a portable cooking-stove, reserves of food and medicine, and a first-aid kit. Amy had also to increase the weight of the flimsy Moth's load with her collection of tools and spares (which included an extra propeller that had to be conveyed externally, lashed to the centre-section struts). Her preparations for the 11,000-mile trip were completed by the acquisition of maps on rollers covering it in varying detail.

One of Amy's goals was to improve on the time taken by the Australian flyer, Bert Hinkler, who had spent fifteen days making the first solo flight from Britain to Australia in 1928. When she took off from Croydon on 5 May 1930, she was in many ways terribly ill-equipped experientially for the gruelling task ahead of her: 'The prospect did not frighten me, because I was so appallingly ignorant that I never realised . . . what I had taken on.'

As everyone knows, Amy succeeded in reaching Darwin. On the way she had constantly to pump petrol from one tank to another, which made her feel sick; she flew through a desert sandstorm and monsoon rain, had to stay awake and keep up her spirits during long periods of poor visibility by reciting nursery rhymes, had to keep calm when her engine started to splutter badly, made one or two rickety landings, and was burnt and blistered by extremely strong sunshine. Arriving in Australia on 24 May meant that she had not beaten or equalled Bert Hinkler's record, but any sense of failure she had about this was soon eclipsed by the wild enthusiasm of the Australians who mobbed, fêted and eulogised her. Congratulations poured in from all over the world; the London *Daily Mail* made her a special award of £10,000; books and articles about her sprang up like mushrooms, and so too did popular songs. As well as the most famous:

> Amy, wonderful Amy,
> How can you blame me for loving you . . .

there were 'Queen of the Air', 'Aeroplane Girl', 'Johnnie, Heroine of the Air' and 'The Lone Dove'.

High spots in Amy's subsequent career were her flight from England to Japan in a DH 80A Puss Moth, *Jason II*, in 1931, in a total flying time of seventy-nine hours, and her meetings and friendship with Amelia Earhart, whom she much admired, and who seemed able to withstand the vicissitudes of being in the spotlight with more balance and serenity than Amy. Amelia encouraged her to stick to her guns and to try to make a serious contribution to aviation without being deflected by gimmicks and publicity stunts.

Amy seemed generally to collapse under the stress of promotional lecture tours, and it was when recuperating from one of these that she met the raffish, hard-drinking, talented and apparently irresistible-to-women aviator, Jim Mollison. After a rapid whirl of parties and socialising, they were married in London in 1932. For some time they were the darlings of the media (though it must have been hard to invest Jim with the same heroic image as Amy), but their life together was far from happy. Amy became more and more depressed by Jim's drinking and womanising; despite her celebrity and veneer of sophistication, she was still essentially a decent, even slightly puritanical, girl. Their problems were possibly compounded by the fact that in their joint aviational endeavours success eluded them. They both continued to set new solo records, but togetherness in the air seemed as dogged with frustration as their efforts to share life on the ground. Amy was frequently near to breakdown, there were money troubles, and she found herself drinking to avoid facing the miserable realities of their relationship. When Jim took off for America in 1935 without consulting Amy, she took up residence in Paris where for a brief period she ran a small air cruise company. By 1936, however, she had abandoned this work, suffered several injuries as the result of air crashes, and separated officially from Mollison.

She had experimented with writing and journalism, but found this stressful rather than fulfilling. By 1937, when Amelia Earhart had disappeared on her round-the-world flight, Amy was beginning to feel that she too had come to the end of her flying career. For a very short time in 1939, however, she worked as a pilot for a ferry company based near the Solent. The outbreak of the Second World War brought this job to an abrupt end, and with some reluctance she joined the women's branch of the Air Transport Auxiliary (ATA) – she considered it demeaning to be employed as a flyer at only £6 a week despite her outstanding experience (male pilots automatically earned far more). She also, not unnaturally, resented having to be interviewed for her suitability for the work as if she were a 'typical CAG [Civil Air Guard] Lyons-waitress type'.

Nevertheless Amy settled into her ATA job and, according to several of the female flyers who worked with her, was at ease, friendly with them all, and without pretentiousness. (She seemed to have overcome the inability to feel part of a group which she had experienced so many years earlier in her office and shop days.) 'She fitted into Hatfield like a hand

in a glove. . . . She settled down simply, to do a job of work,'
wrote Alison King, an ATA operations officer, in her autobiographical book *Golden Wings* (1956).

Jim Mollison was shortly also to become an ATA ferry
pilot, but any contact which he and Amy might have had
through enforced proximity was purely casual – they were
now divorced. Amy showed no signs of entering into any
new romantic relationships. She was busy with the job in
hand, and both her sister, Molly, and Pauline Gower, the
head of the women's section of the ATA, felt that she was
happier than she had been for some time.

One or two of Amy's fellow pilots have commented retrospectively that she hinted at expecting to be killed in an air
accident. She once said that long-distance record-breaking
flyers always 'copped it' eventually, and prophesied that 'I
know where I shall finish up – in the drink.'

This was to be her end, not in some remote ocean during a
long-distance flight but in the waters of the Thames Estuary,
apparently off-course in bad visibility in an Airspeed Oxford, on a ferrying mission. As with Amelia Earhart, there is
a mystery associated with her death. It seems certain that she
baled out: someone was seen parachuting down into the
water, and a flying bag found nearby carried Amy's name.
However, her body was never found; ironically, Lieutenant-
Commander Walter Fletcher, captain of the trawler HMS
Haslemere, who dived in to save her, was also drowned. From
the day of Amy's death until the present time, rumours have
circulated that a *second* parachutist's body was seen in the
estuary. If so, like Amy's, it was never recovered. The
suggestion has been made that – against all the rules of the
ATA – she was carrying an unscheduled passenger, and that
this 'secret' transporting was part of an intelligence mission,
or counter-espionage plot. This seems unlikely, to say the
least. A fellow ATA pilot poured scorn on the suggestion,
commenting, 'Can you imagine someone in MI5 saying
"Let's send one of those women in their bright yellow
trainers" when they had plenty of well-qualified men pilots
available?' Also, can one imagine MI5 using for some secret
mission the woman who had, only a few years earlier, been

just about the most famous Briton in the world? Just as Amelia Earhart's fans will not let her die, it seems that there are those who will go to extraordinary lengths of rumour and speculation to keep alive the name of Amy Johnson.

Their endeavours are superfluous. It was an irony that Amy, the extremely experienced and adept flyer who had caught the imagination of the whole world, should be the first member of the ATA to be killed – but, as a woman who pushed out the boundaries of aviation, who trained herself to attempt and achieve great things, she has secured her niche in history.

Her significance as a symbol is vividly underlined in girls' flying stories throughout the 1930s and subsequent decades. In C. M. Drury's novel *Kit Norris, Schoolgirl Pilot* (1937) the heroine reiterates on several occasions that she wants 'to be a second Amy Johnson'. In fact, her school-friend's father, Air Marshal Lyth, becomes rather fed up with this hero worship ('although he held a high position in the service . . . he could not get used to the idea of women flying'): 'I think many young women will be fired to follow in Amy Johnson's foot-steps, but remember she has set a very high standard, and it will mean real hard work,' mutters the irritated Air Marshal.

However, Kit presses her symbol to good use by consider-ing how her heroine might have acted if confronted by the difficult situation which soon confronts her. She discovers a nest of smugglers, whose lavish loot of contraband watches is stowed in a waiting plane. Amy J., of course, would have taken decisive action – and so does Kit. She temporarily 'kidnaps' the plane and its contents, and decides to fly it to the nearest police station. This isn't quite as easy as she expects: she gets lost, inadvertently flies out over the Channel and eventually lands on a small island which, while fairly near to both England and France, has no telephone or means of communication with the outside world. During her long and hazardous flight she had learned the meaning of loneliness: 'You know, it's being quite alone that stumps me!' Fortunately, the much-admired aviatrix who has blazed so many long and lonely trails, is there once again in an inspira-tional capacity to help Kit cope. She lights a fire on the shore, and with the aid of a blanket and its flames, spells out an SOS in Morse to a small schooner out at sea. She and the haul of watches are, of course, rescued, and Air Marshal Lyth, suitably impressed, promises that he will help her to have flying lessons as soon as she is old enough. Meanwhile she is earnestly urged to stick to the Girl Guides if it is adventure that she is after.

Eileen Marsh wrote several novels with aeronautical themes during the 1930s, as well as some short stories for magazines. Her full-length books include *Two Girls on the Air Trail, Peggy, Parachutist, Air Girls at School,* and *Wings at Midnight.* In *Lorna – Air Pilot* she introduces an unusual mother-and-daughter flying duo. Lorna is the pilot and Mrs Carter specialises in aerial photography. The family have fallen on hard times because Lorna's father, an ex-pilot, is crippled as the result of a motor accident. Fortunately they have managed to hang on to their private aircraft, 'a Heston Phoenix roomy high-wing monoplane' which has plenty of room for the pilot, photographer and 'camera equipment, spare plates and so forth'. Mr Carter used to pilot the plane while his wife made her pictures, so Lorna takes over his role. Her aim is to make enough money – £3000 – to enable her father, in those pre-National Health Service days, to afford the surgery which he needs if he is to walk, and fly, again.

At times it is a little difficult to believe that the Carters are impecunious. Their home seems gracious and their life-style expansive – not many teenagers, after all, can nip up in their own aeroplane whenever they have to work out some knotty problem, but Lorna 'always thinks better in the air'. Naturally mother and daughter make a great success of their flying photography business, thanks largely to an input of ingenious ideas from Lorna. She develops a nose for news and becomes a kind of aerial reporter for prestigious papers such as the London *Daily Wire.* She and her mother rout all sorts of baddies on their way to working up their £3000 fund. They also encounter some nasty anti-feminism from professional rivals: 'You didn't ought to let yourself be beat by a couple of women, even if one of 'em does fly an aeroplane. Can't you try some rough stuff?' (The speaker is, of course, supposed to be American.)

An interesting factor about Eileen Marsh's flying stories is that although they are often exciting and atmospheric, they contain very little about the technicalities of aviation, or a pilot's-eye view of things. One wonders if the author just capitalised on the general airminded mood and slotted herself into the sub-genre, knowing that certain key words and phrases were in themselves sufficient to conjure the appropriate response from readers, without any understanding of aviation or research on the writer's part.

The short story 'Vera Flies Alone' by G. E. Hopcroft, which appeared in an early 1930s edition of *Every Girl's Story Book*, flits in and out of reality. Fifteen-year-old Vera Seabright has 'flown Dad's monoplane a dozen times at least but he still won't let her go up in it alone'. ('Dad' is wealthy Sir William Seabright whose hobby is aviation. The story makes the point several times over that on the whole only affluent girls had opportunities to become pilots in the 1930s.) Vera is on holiday at a South Coast fishing village; she is staying with her friend Alice Stanley (whose father also has his own plane, though Alice hasn't 'ventured to so much as touch the controls'). In a terrible storm at sea, Vera takes off in her friend's father's monoplane to rescue a seaman whose ship has broken up on the notorious Flat, a submerged bank of shingle and sand. The lifeboat cannot get through, so everything is up to Vera. She not only has to land on the Flat but to take off from it, with the tide already rising again, making her runway shorter by the minute:

> At last – oh, glad moment – the engine began to roar, and the aeroplane rushed forward. Would it dash into the sea, or would it rise? For one awful moment it seemed that the wheels of the under-carriage would enter the water before the machine could lift, but, just as Vera, her heart in her mouth . . . was steeling herself for the plunge, the monoplane rose clear of the Flat and went soaring landwards, its engine roaring as if in triumph.

The whole village turns out to cheer her, and the moral, of

course, is that as many teenage girls as possible – especially those like Alice whose fathers own aeroplanes – should learn to fly. An interesting social reflection is that the solitary, stranded sailorman (after a horrendously cold and wet night spent clinging to the wreckage of his craft) still has enough energy to 'gasp' in amazement when he sees a female climb out of the plane and to comment copiously though appreciatively on this: 'T'aint many girls as would arisked their lives, in a manner o' speaking, for the sake o' a poor sailorman.'

It seemed that in spite of the achievements of Amy Johnson and an ever-growing number of women flyers, doubts and prejudices about their capacities were still firmly rooted in many people's minds. They considered that there was just something not quite respectable about the whole business of girls piloting planes. It was, perhaps, too potent a symbol of women controlling their own destinies.

A rather sad reflection of the struggles which some girls might have before they could become airborne appeared from time to time in the shape of a quarter-page advertisement in *Popular Flying*, the aviation monthly edited by W. E. Johns, creator of the redoubtable Biggles. It shows a very small, bespectacled girl saying

> I wish I were an angel, mummy,
> I would so love to fly . . .

[while Mum replies]

> Hold your tongue, precocious child,
> You may be by and by.

The doggerel continues, with the little girl asking why she can't be like Daddy who could 'gratify his whim' [and] 'Learn at 40 bob an hour/In aeroplanes at Lympne'.

One is probably never too young to make a feminist point. Things would certainly have gone better for the adult heroine of Gilbert Frankau's 'The Scarlet Parachute' (in the 1937 *Christmas Pie*) had she stood up for herself rather more sturdily. Christina is married to Pat Paget; they are both pilots who perform a vaguely romantic double act in their planes *Hearts United* and *Ever Faithful*. Christina is, we gather, amply proportioned (her charming husband refers to her in private and in public as 'the big lump'), with fairish, non-descript hair and an undistinguished face. Pat has the looks of a Hollywood film star, and a lot of Irish blarney to go with them. He calls the tune in the relationship (though it is not clear that the narrative view sees it that way), and Christina fears that Poppy Delafield, the rich and stylish daughter of a visiting dignitary, is succeeding in making Pat fancy her: 'Poppy was a most alluring girl, *petite*, with the most glorious dark hair, and the loveliest hands. Such lovely, such useless hands, never smeared with grease, never roughened, never torn like Christina's!'

Christina has another problem; she is beginning to lose her nerve in their aerobatic displays, and if she can't be Pat's professional partner, she fears that there is little left for her. Her doctor confirms that her flying days are numbered: 'I admit you're airminded. . . . But you're a woman, and you're not an ordinary commercial pilot. Take my advice, and give up this stunt-flying.'

Poppy and Pat flirt openly, and when Christina takes her husband to task about this he is insultingly dismissive. She decides that she'd rather see him dead – or die with him – than give him up to Poppy, and devises an extremely compli-cated test of his faithfulness. If he doesn't come up to scratch, she will see that when he makes his parachute drop – the climax of their act – he has a faulty 'chute and will fall to his death.

Pat fails Christina's love test; she watches him jump in the damaged parachute and goes into a paroxysm of screaming, coming out of it only when she hears Pat's voice saying, 'Big lump, my own darling big lump, what's the matter?' It is all a horrendous dream – and the normally insensitive Pat turns out to be quite understanding as she sobs in his arms that she will have to give up flying. He even – nobly – agrees to take a few days off from it himself.

As long as flying fiction for women and girls was used as a mere peg on which to hang romance or banal adventure, it would fail to provide genuine excitement or satisfaction. In

the early 1930s, boys were fortunate in having aviation stories written by men like W. E. Johns and George Rochester
who, however much they might be criticised on literary or
social counts, knew everything about flying from the inside
out. Girls had to wait until later in the decade to be as well
served.

CHAPTER SEVEN

The Elegant Enigma

After all, when you spend your time flying, you meet so many jolly people, and get so many good friends, that there doesn't seem time for sweethearting.

DOROTHY CARTER,
Star of the Air (1940)

DESPITE the heroic aura which so often surrounded her, Amy Johnson seemed to some extent to be a derivative of the traditional 'girl next door'. Despite her flaws – perhaps even because of them – she remained an extremely sympathetic character. Her contemporary and rival, Jean Batten, was not, although for many years this was obscured by the brilliance of her aviational performances, by the fact that she always looked absolutely wonderful, and by the production of a cleverly whitewashed autobiography which for decades was taken as definitive. Jean Batten, the dentist's daughter from Rotorua, New Zealand, always knew what she wanted, and got it. Determined to the point of ruthlessness, she was exploitive, both emotionally and financially, and, apart from the obsessive lifelong affection which she felt for her mother, sustained no close relationships. She ranks with Amelia Earhart and Amy Johnson as one of the most celebrated aviatrices of the inter-war years, and, as with them, a mystery was to surround her death.

Jane (later known as 'Jean') Gardner Batten was born in September 1909, only a few weeks after Louis Blériot's first flight across the Channel. Her ambitious mother had pinned a newspaper picture of Blériot and his plane to the wall by her baby's cot, to signify that the next generation could continue the business of making history. It seems unlikely, however, that Jean could have been much influenced by this.

Jean's parents, Fred and Ellen, had hoped that she would become a concert pianist; music and dancing played a big part in her childhood, but she was more interested in geography and travel, and, at nineteen, decided to emulate the exploits of her childhood heroes – the aviators Alcock and Whitten-Brown and the Smith brothers. Her ambition was strengthened by a meeting with the Australian flyer, Charles Kingsford-Smith, who took her over the Blue Mountains on her first flight when she was on holiday in Australia. Fred Batten, with memories of First World War flying accidents, and on the grounds of expense, was opposed to his daughter taking up aviation, but with Jean and Ellen presenting a united front, he soon gave way. Symbolically, the family piano was sold to help with the fees for Jean's flying lessons, and mother and daughter came to England, which they considered to be the aviational centre of the world.

Training at Stag Lane, Jean was delighted to meet other aviatrices at the London Aeroplane Club, including Amy Johnson, whose records she made up her mind to beat. Once she had achieved her 'A' (private pilot's) licence, she expected to find sponsorship for a flight from England to Australia, but, because of her lack of experience, there were no takers. So Jean worked to obtain her commercial licence, by which time money was running out and she had to pawn many personal possessions to cover the cost of the 100 hours of solo flying that the 'B' licence demanded.

From then on Jean had little difficulty in obtaining sponsors for her long-distance flights. Help was sometimes given altruistically by Lord Wakefield and other benefactors with a general interest in aviation: it was often obtained from one or other of Jean's male admirers. (She encouraged more than one of these, at different times, to believe themselves engaged to her. Some were rich, and able to shrug off the loss of large sums of their money, even if they had originally thought of subsidies for Jean as loans rather than gifts. Others, like Fred Truman, a New Zealand RAF pilot who presented her with all his savings, were unable to contemplate with equanimity not only the loss of her affections but the loss of their money.)

After a couple of abortive and expensive false starts (one of which ended in Karachi and the other in Rome), Jean flew solo from Lympne in Kent to Darwin in Australia, completing the journey in fourteen days and twenty-two-and-a-half hours, and spectacularly beating Amy Johnson's 1930 time by over four days. Less than a year later she made the return trip, thus becoming the first woman ever to fly solo from England to Australia and back.

Also in 1935 she flew solo from Britain to Brazil, setting a world absolute record (for pilots of either sex) and becoming the first woman to fly this route solo. In 1936 she made another world absolute record, flying solo from England to New Zealand in eleven days and forty-five minutes, and, to cap everything, in 1937 flew again from Australia to England to set yet another solo record (for either sex) of five days eighteen-and-a-half hours.

Ian Mackersey's intriguing and investigative biography, *Jean Batten: The Garbo of the Skies* (1990), has thrown light on many facts which Jean and her mother rigorously kept secret. Although enjoying mass adulation and media exposure for brief periods, Jean and Ellen spent most of their life together reclusively sealed off in an intense mutual affection. They kept everyone else – even Jean's father and brothers – at arm's length. The reserved and modest behaviour which Jean rather engagingly described in her autobiography – *My Life* (1938), reissued as *Alone in the Sky* (1979) – seems actually to have been an indication of deep sexual repression. Aptly dubbed by Mackersey 'The Garbo of the Skies', she carried her 'touch me not' policy to extreme lengths, even wincing if someone gently took her arm to guide her through crowds, often failing to acknowledge acquaintances, and treating friends (to whom she owed favours) outrageously if they presumed to talk to her with simple, human directness. Like Garbo, by separating herself from her peers she gave credence to legends of her own super-capacities (her physical beauty was self-evident, and unneedful of mystique); at a fairly basic level, for example, she implied a kind of superiority by claiming that her routine of strict bladder-training enabled her to manage for nine-hour stretches without access to toilet facilities in her Moths or Gulls: 'I cut down my fluid intake to camel-like levels.' In fact she had a top-secret in-flight toilet installed.

Jean's career flourished during the romantic and myth-making period of aviation, and there is little doubt that she was more skilled than many other celebrated pilots of the time. She was saved from death on more than one occasion by the brilliance of her navigation and technical control, but, although she could negotiate her fragile craft through extremes of adverse climatic conditions, she seemed unable to balance or control her own life.

She gave up flying at the start of the Second World War and offered herself and her Percival Gull to the Air Ministry in London, expectedly for use in the ATA. To her chagrin, the aeroplane was accepted, but not the aviatrix. Surprisingly, her rejection was on the grounds of 'poor near-ocular vision', but it seems more likely that some people in high places doubted her ability to work as part of a group, or even her loyalty to the Allied cause. (Her name had been linked with some possibly pro-Nazi Swedes, and, in a recent dash across Germany to get back to England, special travel facilities had been sought from and granted by the Germans.) For a while she persisted in trying to obtain work as a ferry pilot: eventually she gave up and undertook other forms of war service. She drove an ambulance, and then worked on the production line at the Royal Ordnance factory at Poole, inspecting the Hispano Suiza 20-mm cannon used in the Spitfire. She also travelled extensively throughout Britain to speak in support of the National Savings campaign, surely proving her patriotism beyond doubt.

After the war Jean and Ellen went to live in Jamaica, and later in Tenerife. Once again they were sealed into their secret, intense relationship. Far from resenting the domination which Ellen always exerted over her, Jean seems to have needed it. When her mother died at nearly ninety, she was distraught, and, although fifty-seven years old, never came to terms with her loss. She continued to resist overtures of friendship from various people (though there are indications that she toyed with the possibility of one or two hopefully romantic relationships), but made a 'come-back' of a limited nature into public life. Mini-skirted, high-booted, with her face surgically lifted and her grey hair dyed startlingly brunette, she looked, at sixty, like a woman in her early forties. (Later on she was to experiment with both auburn and blonde coiffures.) Her involvement in several prestigious aviational events continued spasmodically until the early 1980s.

Then she suddenly became 'a missing person'. It was thought that she had gone to Majorca, but she gave no address to anyone. Her passport expired; mail piled up; her bank had no record of financial transactions; her publishers lost contact, and neither the Foreign Office nor the New Zealand High Commission could trace her. The secrecy which she had so often wrapped around herself now seemed totally and permanently engulfing.

Ian Mackersey and his wife Caroline decided to make a television documentary feature on Jean Batten, and their extraordinarily persistent researches ultimately pieced together the story of the last months of her life. She had settled at Palma, Majorca, in a modest hotel, where she lived a reclusive life. The hotel staff felt that she must be poor, because she had so few possessions and wore clothes which were obviously far from new and had sometimes been repaired. In 1987, when the Mackerseys eventually made contact with them, they still remembered Jean as a very beautiful woman, but had no idea of her former celebrity. In her early seventies she was still vigorous and took strenuous walks every day. On one of these her leg was bitten by a dog, and Maria, the hotel maid, on seeing the teeth marks, suggested that a doctor should be called. Jean refused, and seemed to feel that she could heal herself (Ellen had always strongly supported theories of natural healing); she said she had no fever, but the maid noticed that after some days she stopped eating, and drank great quantities of water. On 22 November 1982 Maria came in to clean the room, and Jean, fully dressed but weak, drank some soda water. Maria went to get clean sheets, and when she returned Jean was dead.

In retrospect it was a mystery that neither the Batten family nor the New Zealand government was informed of her death. Apparently, however, the Majorcan authorities had sent formal notification to 'the New Zealand Embassy in Madrid' – but no such embassy existed, diplomatic relations with Spain being handled by the New Zealand Embassy in Paris. The message that could not be delivered in Madrid was returned to Majorca and filed away in the archives. Consequently Jean Batten, the one-time heroine of hundreds of thousands of people, was buried in a pauper's grave (despite the fact that she had an estate of approximately £100,000).

Because, in such communal graves, coffins were piled on coffins, often disintegrating, it was suggested that any attempt to disinter Jean's remains and return them to New Zealand would be impossible. Local officials tactfully pointed out that, with the passage of time, name labels fell off coffins and the bones of different bodies could become mixed up. There was talk of a memorial which would fittingly commemorate her and the honours which she had brought to New Zealand. At the end of 1989, a cabinet minister attended the unveiling of a plaque on the wall above the communal grave; it was a low-key affair, but perhaps, in an ironic way, appropriate for the Garbo of the Skies.

In her heyday, the popular image of Jean Batten certainly influenced writers of flying stories for girls. She was even made one of the patrons of the London-based Woman's Book Club, in company with distinguished authors such as Vera Brittain, E. M. Delafield, Daphne du Maurier, Naomi Jacob and the Baroness Orczy.

Dorothy Carter's books first began to appear in 1936 with *Flying Dawn*. Little is known about the author, except that she learned to fly in a British Aircraft Co. Swallow (actually the German Klemm L25 angular two-seat monoplane), which could fly hands-off, 'showing how safe modern flying is'. This information is conveyed in an article entitled 'Flying as a Career for Girls' which Dorothy wrote for the June 1938 *Girl's Own Paper*. It warmly praises the record-breaking Jean Batten and the air-taxi pilots, Pauline Gower and Dorothy

Spicer (see chapter 9), for the opportunities which their aviational exploits opened up.

Efforts to find out more about Dorothy Carter, as a pilot and as a writer, have proved abortive – but she was certainly one of the first authors to create really convincing flying stories for girls. There is an indication that she was very aware of Captain W. E. Johns's Biggles books, which might possibly have stimulated her to create female equivalents: at the end of her third novel, *Wings in Revolt* (1939), she introduces a musical-comedy type of character – a sea captain who indulges in some arm-wrestling with a US consul. His name is Captain Bigglesworthy – a nifty dig, surely, at the redoubtable Captain Johns and his hero, James Higglesworthy.

Dorothy Carter's second novel, *Mistress of the Air*, was originally serialised from October 1937 in the monthly *Girl's Own Paper* and later published in book form. It introduced Marise Duncan, a new flying heroine, who was authentic and appealing enough to inspire six novels in all. The first three books in the series, *Mistress of the Air* (1939), *Star of the Air* (1940) and *Snow-Queen of the Air* (1940), have peacetime backgrounds. The last three, *Sword of the Air* (1941), *Comrades of the Air* (1942) and *Marise Flies South* (1944), are set respectively in Germany, Russia and the Far East during the Second World War (see chapters 8 and 9).

In *Mistress of the Air* Marise shows her mettle in the first few lines. She is 'the youngest girl ever to compete in the King's Cup Air Race' and although her airline pilot father, Captain Duncan, is trying not to pressurise her, she realises that she must make even greater efforts: 'Second prize isn't good enough,' said the girl quietly. 'I'm going to catch that Hawk, Daddy. We might get it on one of the bends.'

Their Vega Gull roars on, taking Marise and her father to victory. The course of the race, as described in the book, from Hatfield to Sacombe and Hoo End, is accurate, and the idea of a female winning this highly regarded event was also based on fact. In 1930 Winifred Brown, who as well as being a flyer, was a yachtswoman and an ice-hockey player, had achieved the trophy in her Avro Avian biplane. She was the first girl ever to win the race, and soon afterwards Amy

Johnson and other aviatrices were shocked to find that she was appearing at the London Coliseum in flying kit, with her winning aeroplane on the stage beside her. The publicity given to Winifred's variety turn was particularly distressing for female flyers who struggled constantly to find serious rather than gimmicky jobs in aviation.

There was no danger that the determined Marise would take on any stupid jobs after winning the race. Mrs Duncan, her mother, who throughout the saga is the mouthpiece of conservatism and unconscious anti-feminism, has little appreciation of Marise's achievement:

> 'Very nice, dear,' said Mrs Duncan when she was shown the cup. 'But is it going to help you to get a post? Flying is all very well for fun, but I really think you had better make up your mind to find a job as a secretary, or something . . .'

Even when Marise is offered work as an instructress at a flying school in Kent, Mrs Duncan is still depressingly harping on about shorthand-typing: 'It's more – more respectable,' she says petulantly.

Nevertheless, Marise takes on the job at the flying school. Her boss, 'Tiny' Cator, explains that most of her pupils will be female, as lots of women and girls now have ambitions to fly. He takes to Marise straight away, but is insistent that she should upgrade her image:

> 'I want you to rig yourself up like a young lady flier as shown on the films . . . the public expect it, and no girl learning to fly is going to have the least bit of confidence in you unless you are wearing a smart set of white overalls, looking like Myrna Loy.'

He appears to be referring to the popular film *Wings in the Dark*, which starred Myrna Loy and Cary Grant. Publicity shots from it confirm that the movie star was one of the few women who looked really beautiful in a flying-helmet – Jean Batten, of course, was another. Both favoured white helmets and overalls, when possible, and Marise appears to have been pleased to emulate them.

The flying club at which Marise works is the Bonnington, near Lympne. Dorothy Carter must have known the area well (Lympne was of course an aerodrome in the 1930s) and she adds authenticity to her narrative by using real place-names: Littlestone, Aldington, and even Smeeth, which sounds extremely fictitious but *does* exist. The main story-line of *Mistress of the Air* focuses on an aerial expedition to the Ellice Islands (now Tuvalu) in search of an Elizabethan pirate's treasure. The search takes a long time to get started, partly because Marise has first to train the three treasure-hunters to fly, and partly because the author seems so attracted to the world of the 1930s' flying clubs that she is reluctant to move the action away from Bonnington.

Marise's three pupils are Pauline Williams, a girl of her own age; her grandmother, the feminist tartar, Lady Wilhelmina Williams, whose determination is close to obstinacy ('I'd sooner sing the 'Red Flag' in a Fascist headquarters than tell her she can't do anything she's set her heart on doing'); and Lady Wilhelmina's friend, the recently retired General Barclay, who is at first furious at being given flying lessons by 'a chit of a girl'. He snorts contemptuously that women's place is in the home, and that they should stop pushing themselves into aviation where 'they're outclassed by men all the time'. Marise, of course, has the satisfaction of putting him right: 'I won the King's Cup Race, anyway, against a lot of men.'

He soon responds to Marise's expertise; she is put in charge of the expedition to the Pacific, and, once again, Dorothy Carter's detailed knowledge of aircraft becomes evident when her heroine recommends the Saro Cloud as the most suitable machine for the job. As Squadron Leader Dennis L. Bird commented in an assessment of Dorothy Carter's stories (in *Collector's Digest*, 533):

> This is an excellent choice – a twin-engined eight-seat flying-boat with retractable wheels, an amphibian in fact. . . . The author really knows her stuff. Passing remarks about aircraft handling are authentic, and so are the types mentioned: Avro Tutor, General Aircraft Monospar, 'a little Tipsy monoplane'

(Belgian), the Sikorsky 'China Clipper' at the Azores. Names to send a thrill down the spine of the aviation enthusiast.

Dorothy Carter carried girls' flying stories a long way forward; she had a flair for conveying technical detail with a light touch, so that even girls without special aptitude or interest could find it absorbing:

> Turning into the wind, Marise opened the throttle full out for the take-off. As they reached the middle of the lagoon the flying boat came off the water and soared into the air. 'I must get that tear [in the wing] mended at Honolulu,' said Marise mechanically.

In *Star of the Air* Marise goes to Hollywood to perform the flying sequences in an ambitious and extremely colourful film. Mrs Duncan, of course, protests that 'It isn't respectable', and Marise is moved to respond: 'Mum, you can't say Shirley Temple isn't respectable.' The high spot of this story is Marise's acquaintance with the stunt pilot Jim Grant, who took part in aerial actions during the First World War, as a result of which he has only one eye, and a limp. Despite all this, he has a rugged charm. Marise begins to like him enormously, and as he crops up in several of the books, readers start to wonder about a possible romance between them – despite the age difference. Nothing develops, however. Like W. E. Johns's Worrals of the WAAF, Marise remains firmly virginal throughout all her exploits, and, in the last book of the series, she restates her attitude: 'I'm quite used to boys – but only as colleagues in the air.'

Marise is certainly more fortunate than most real-life aviatrices in achieving all sorts of fascinating jobs. *Snow-Queen of the Air* takes her to Canada, to the snowy wastes of the Arctic where she helps her father, who is acting as consultant for a new northern airline route. The story is remarkably atmospheric. Marise has to learn to survive in the Arctic conditions, to land on and take off from cracking ice, to fly through blizzards, to adapt to flying off skis, and to drain the oil from her engine as soon as she is grounded so that it doesn't freeze. In a Rider Haggard-type episode, she has to make a forced landing among an Indian tribe who have a long-held legend that their future queen will drop out of the skies. Marise does, of course, and becomes – fortunately for only a brief period – their 'Snow-Queen of the Air'.

As well as writing her novels, Dorothy Carter produced five short stories, mostly with flying themes, for the *Girl's Own Paper* during 1939 and 1940. These were 'Edna: Night Watcher' (May 1939), 'May's Monoplane' (June 1939), 'Lizzie of the Bush' (August 1939) 'Patricia's Party' (September 1939) and 'Sally's Solo' (February 1940). The most exciting of these was 'Lizzie of the Bush', in which fun-loving Elizabeth, who swims, sails, plays a good game of golf and pilots her own aeroplane, heroically saves whole flocks of sheep during a bush fire by ferrying them out of danger in a De Havilland Dragon. To control this in the face of 'currents and turmoils' which throw it about like 'paper in a high wind', Lizzie has to maintain a vice-like grip on the wheel of the control column and brace her feet firmly against the rudder bar. Neither Amy Johnson nor Jean Batten could have done better.

CHAPTER EIGHT

Aviatrices in Australia and Africa

TO Joan's amusement one of the pygmies presented her solemnly with a tiny shield and spear. Joan patted him on the head, and thereafter he was the proudest pygmy in the tribe !

(Illustration from 'The flying Schoolgirls', *Schoolgirl,* 1933).

Being alone in an aeroplane for even so short a time as a night and a day . . . with nothing to observe but your instruments and your own hands in semi-darkness, nothing to contemplate but the size of your small courage . . . such an experience can be as startling as the first awareness of a stranger walking by your side at night. You are the stranger.

BERYL MARKHAM,
West With the Night (1942)

AMELIA EARHART, Amy Johnson and Jean Batten seemed in their individual ways temperamentally well suited to cope with the loneliness of the long-distance flyer. Partly from volition and partly from necessity, their main aeronautical activities centred on the breaking and setting of speed, distance and endurance records.

Two different kinds of long-distance aviation, 'mercy missions' and the transportation of freight and passengers, were to become the chief functions of the Australian flyer, Nancy de Low Bird. (The 'de Low', which she dropped for professional purposes, came from her mother's partly French side of the family: so too did her Christian name, for she was called Nancy after the capital of Lorraine.)

The appellation of 'Nancy Bird' suited her well. 'Bird' suggested not only her passion for flight but also her diminutiveness. She comments in her 1961 autobiography, *Born to Fly*, that, at seventeen, when she started to have flying lessons, 'If I stood very straight indeed I was under 5 feet 2 inches high.' Cushions had always to be placed behind and beneath her so that she could reach the rudder bar, and these were to become 'the bane' of her flying life.

Nancy was born in the sawmilling town of Kew in New South Wales. With her freckled face, red hair and short but stockily determined stature, she seemed at thirteen, when she first started craving to fly, rather like a story-book heroine. The second daughter of a family of six, she had by then left school, where she liked nothing but sport, to keep house for her widowed father, a country storekeeper.

Every penny that she could spare from the allowance paid to her by her father was put aside for flying instruction, and every minute that could be snatched from her household duties was spent in studying Swoffer's *Learning to Fly*, a manual which she had ordered by post from a Sydney bookseller, and every page of which she devoured again and again. However, later in life she reflected that this was definitely 'a wrong way to begin a flying career . . . One thing I am certain about is that you can't learn flying from a book – you can only learn it in the air.' She likened book study of aviation to the reading of medical journals by the untrained: 'You end up with a lot of pains that you never had before.' Flying problems could be magnified out of all proportion if pupils encountered them only in books without the back-up of at least one or two elementary practical lessons.

Of course, even if at thirteen Nancy *had* recognised how imperative it was to have flying lessons from the beginning, her youth would still have debarred her from taking these. So too would her lack of funds, but by the time she was seventeen she had saved the £200 for the course of instruction for her 'A' licence. She records that on 11 August 1933 she had her first lesson in a DH60 Cirrus Moth from 'that great airman Charles Kingsford-Smith' at Mascot aerodrome (later to be developed into Sydney airport). It was a tremendous thrill for the teenage Nancy to receive lessons from this celebrated pilot who had made round-the-world and Australian record flights, but she found him to be a less than brilliant instructor and was happier when her tutelage was transferred to Pat Hall, 'a born teacher'.

Kingsford-Smith had made it clear to Nancy that he didn't approve of women in aviation: 'It's not the right place for them!' Nevertheless it was he who probably gave her the

happiest moment of her young life when, after her performance at her first flying lesson he pronounced, 'Not bad for a
first try: you'll learn.' The elation caused by this remark, and
her own satisfaction at becoming airborne after waiting so
long to fulfil her ruling passion, gave Nancy a tremendous
and sustained influx of energy. Although she spent only
twenty minutes a day in the air, she went out to Mascot every
morning and remained there until flying was stopped for the
day by failing light. After her mile-and-a-half walk from the
tram stop to the hangar and the red dirt runways, she was
ready to learn about everything to do with aircraft. At first
none of the mechanics took her seriously ('What the devil
does she want to learn engineering for?'), but once Tommy
Pethybridge, Kingsford-Smith's chief engineer, became convinced that Nancy was no mere fair-weather flying freak, he
taught her how 'to grind valves, how to chip carbon off
pistons, how piston rings should fit, how to time a magneto
and check tappet clearances'. Once Nancy had acquired her
private licence in 1933, she began to work for the commercial one which would enable her to carry passengers for 'hire
and reward'. This involved, as well as 100 solo flying hours,
examinations in navigation, engines, air frames and meteor-
ology. The engineers and mechanics at Mascot had by now
decided that Nancy was really seriously wanting a flying
career, and they gave her 'wonderful help' so that she could
overcome the 'awful handicap' of her non-existent mathe-
matics.

However, several people were less than encouraging. Even
at Mascot, there were those who 'shook their heads and said,
"You can't do it, Nancy."' She almost began to agree with
Kingsford-Smith that there was 'no room for women in
aviation . . . men were finding it a hard enough struggle'
without any female competition. The anti-feminist trump
card was always that 'nobody in the world wanted to fly with a
woman' – an argument which remained in use even long
after women pilots were proving their worth in civil airlines
(see chapter 12).

On the positive side, embarking upon her commercial
licence and training with seven other pilots, Nancy became

very friendly with the only other female student, Peggy McKillop, and shortly afterwards went into partnership with her. At nineteen, Nancy was the youngest girl in the British Empire to qualify for a commercial licence. This gave her some prominence, but still no chance of landing a flying job with a commercial airline. She decided that she and Peggy could only make a living out of aviation if they had their own aeroplane for barnstorming around the country. (Of course male pilots warned them that the days of aeronautical displays were over, that two girls who could offer rather less than Kingsford-Smith's and Harold Durant's mock air-battles, or Jim Mollison's and Hereward de Havilland's near mid-air collisions, had no chance of pulling in the money-paying public. However, neither the Little Bird nor the Big Bird – as Nancy and Peggy were known – was deterred.)

Nancy's father, who had never begun to understand why she wanted to fly, suddenly became extremely supportive. He produced £400 towards the purchase of a plane. This, supplemented by £200 from an aunt, enabled Nancy to buy and rebuild an elderly Gipsy Moth (which had come out from England in 1932 with one of those aristocratic avia-trices, Lady Chaytor). By this time Nancy was at home with a variety of planes. During her commercial licence-training she had flown an Avian, a Curtiss Jenny, Puss and Gipsy Moths and other machines.

Like so many of the earlier women flyers, Nancy caused a minor sensation in her choice of clothing. When she began to fly in 1933, 'Nobody wore slacks or even those awful beach-pyjamas' (which were briefly in vogue a year or two later). The 'fussy-looking' dresses of the early 1930s were hardly suitable for climbing into open-cockpit planes 'behind a pro-peller that had already been started up by one of the mech-anics'. The tiny Nancy had the added sartorial complexity of manoeuvring into place on and in front of her build-up of cushions. She took to knickerbockers, which were probably more *outré* for women in the 1930s than in the 1900s and 1910s, when memories of girls in bloomers were still vivid. With her knickerbockers (linen for summer and tweed for winter) Nancy wore long socks and flat-heeled shoes, and rolled her hair into a bun. Making her long ferry, tram and 'shanks pony' journey to the aerodrome each day, she knew that she 'raised some laughs'. When Nancy and Peggy started barnstorming, they undertook a three-month tour, staying at 'two-bob hotels' with a shared room, breakfast, lunch and dinner inclusive. They just about made enough money to feed themselves and to maintain the Moth. Two further tours followed. As well as coping with stunting and joy rides and occasional mechanical hazards, they also had to look good. (It was felt by some people that the success of their barnstorming tours owed a lot to their glamour as female pilots.) With room on the plane for only two schoolchild's-size suitcases, Nancy recalls her wardrobe for a three-month tour:

> I can remember I had two changes of underwear, two pairs of stockings, a flying-suit, a couple of blouses, a leather coat, and, for evening, a long skirt of some crinkly material which didn't crush and a lace overblouse which I had painstakingly embroidered with tiny gold beads.

(Apparently everyone on the big stations at that time used to dress for dinner each night.)

With help from Tom Perry, the president of the Aero Club, who often assisted young flyers, Nancy was able to acquire a larger plane, a Leopard Moth, which enabled her to undertake charter work and become increasingly familiar with the locations of small outback towns and centres and their communications problems. The 'typical-to-the-point-of-caricature' town of Nevertire always intrigued her, with its twenty or so iron-roofed houses, its pub, post-office, garage and two or three stores. It also had green pepper-trees 'throwing their soft shade on the powdery dust of the road', with frequent strong winds blowing.

It was for communities even more remote than this that Nancy started to organise medical flights. The Reverend Stanley Drummond had established the Far West Children's Health Scheme after becoming appalled by the number of children in the outback threatened with blindness from tra-choma, or 'sandy blight'. When Nancy met him, he had established baby clinics all over Australia but, if these were not accessible by rail or road, hundreds of mothers still had to rear their infants with no one to give medical help or advice, even in an emergency.

Mercy flights were the obvious answer and Nancy began to provide these. She started by ferrying a former First World War army nurse to appropriately isolated locations, where the supposed landing-fields often turned out to be small, rabbit-holed paddocks. Nancy worked under contract for the Far West Children's Health Scheme, keeping her plane available for medical tours, but also taking the charter work which provided her bread and butter and helped to pay off the debt still owing on her Leopard Moth.

After her plane had been standing on the ground, the temperature in its cabin was often recorded at over 100°F. There were problems of drought and ubiquitous floating dust; then of hopeless-for-flying visibility, and of airfields and landing-strips becoming lakes. There were also the hazards of sandflies and mosquitoes: 'I've used as much as a bottle of citronella in one day trying to protect my swollen legs and arms.'

There is no doubt that in the 1930s it demanded a special kind of courage to fly alone over thousands of miles of outback, with no radio and frequently no one to check if destinations had been safely reached. In vast waterless areas, chances of survival were few if pilots came down, but nothing deterred the indomitable Nancy. Her work presaged the Royal Flying Doctor Service, which was later to be established largely through what Nancy called the 'God-given vision' of John Flynn. When the Far West Children's Health Scheme could no longer afford to pay her retainer, Nancy continued with chartering and private ambulance flying to bring doctors and patients together.

In 1936 she had 'magnificent fun' flying in the Brisbane to Adelaide air race, in which she won the Ladies' Trophy – by one minute. A 1930s' film dubbed her the Angel of the Outback, and she became celebrated in popular songs and poems about 'Australia's bravest girl'. Eventually, however, she had to give up her medical flights because of extreme exhaustion: 'I felt I would almost rather crash the plane than go on flying it.'

She had, without realising it, also become well known outside Australia. She was presented at Court in London and, in America, flew with Jacqueline Cochran on the maiden flight of the Douglas DC4. In Australia she was appointed Commandant of the Women's Flying Club (later the Women's Air Training Corps).

Nancy married Charles Walton in 1939. After the outbreak of war she was invited by Pauline Gower, head of the women's branch of the ATA in England, to join up with them. Pilots of Nancy's calibre were desperately needed to ferry planes from factories to airfields. She felt unable to do so because of her family responsibilities, but tried to help by compiling a list of Australian women flyers who might be recruited. After the war she was invited to take part in the American Powder Puff Derby, the annual cross-country race organised by the prestigious US women flyers' group, the Ninety-Nines (see chapter 5). In a Cessna 172 with an American co-pilot, Irish Critchell, Nancy came fifth in this 2177-mile race 'across mountains and desert at the height of summer'. She had first to learn to fly the Cessna: 'I'd never flown a modern type aircraft – one with flaps, a wheel instead of a joystick, a right-hand throttle, toebrakes and radio.'

While she was flying frequently across the outback, Nancy had a bizarre near-accident on the ground. The throttle was left open while she went to swing the propeller, the plane ran off and she had to sprint after it to prevent its crashing into some trees. This real-life exploit is echoed – and horrendously intensified – in *Solo for Several Players*, a flying story of 1961 by the Australian writer and journalist Barbara Jefferis. Reversing the dramatic yet cosy mood of the traditional flying doctor stories, from the first paragraph this tale thrusts the heroine into the ultimate aviational nightmare, and brilliantly sustains the suspense until the end. Janet Osborne, who has an unnatural fear of flying, finds herself in a terrifying situation:

Because the take-off was always a torture to be endured with

closed eyes, the plane was a hundred feet or more in the
air . . . before she knew that anything was wrong. She
opened her eyes as the plane began to bank, saw the sheds
and stockyards rising gently over a wing-tip in their familiar
way, turned her head towards Dick to smile the smile that
always amused him because it was so obviously relief at still
being alive, and screamed so that the sound hurt her own
ears when she saw the empty seat.

How her grisly predicament has been brought about is later
explained by Dick Garnett, who got out of the plane (a Piper
Tri-pacer) before take-off 'to chase some horses off the
strip . . . my foot slipped on the step, and somehow I must
have kicked the throttle open as I fell'. Grounded and frantic, Dick hears the plane's motor diminishing relentlessly
towards the north; he knows that the nearest neighbour is
twenty-seven miles away and that there is no possible clear
landing ground more than 'pocket handkerchief size' in the
densely timbered area ahead. Meanwhile Janet, who is in
almost a trance of panic, and fighting nausea with every
lateral movement of the plane, knows just about enough of
the technicalities of flight to pull the joystick up to increase
height, to push it forward to come lower, and – wildly – to
the left or right to keep the plane, as far as possible, in a large
circle not too far from her Brinalli Downs starting-point.

A sharp contrast is made between Janet's lonely anguish
aloft and the feverish activity on the ground. This eventually
involves many members of the local sheep-farming commu-
nity who turn out at first through curiosity ('because some
Sheila's in a plane and can't fly it') but stay to give psychological or physical support as they become caught up in the
drama.

Dick decides to try to bring Janet's plane down at the
nearest aerodrome, Weeringbrinalli, where a doctor, a fire-
truck, an ambulance and all the paraphernalia needed after
a crash-landing are rapidly assembled. He calls up the Flying Doctor base which manages to hook him into radio
contact with Janet so that he can try to talk her down. It is, of
course, a race against time. Janet's hold on reality becomes

increasingly slender: in her terrified view, a quick and total crash is preferable to the slow agony of remaining trapped in the air and speculating about the horrors which seem inevitably to lie ahead. The narrative harnesses Dick's coaxing, wheedling, bullying how-to-fly guidance, and Janet's alternate acceptance and rejection of this, into a masterly build-up of suspense.

She discovers that worse than the fear of flying is the fear of landing, and the climax comes when Dick has talked her into position over the aerodrome for coming down:

> She was glad of . . . any hold-ups, any circumstances at all which could delay for her the moment of attempting to land the plane. She was tired, more tired than she could ever remember having been before in her life, and even the insistent, aching tiredness was useful to her – since it helped her to push to the back of her mind the dread of what lay ahead. Just to go on circling in the quiet air was enough – all that anyone could expect of her.

She knows that ahead of her is the moment she has always dreaded – the moment 'when her own life would depend entirely on herself'. And, of course, she relives her past inadequacies; her feelings that she can never be trusted in an emergency, because some failure of determination 'would make her deny that extra ounce of purpose that made people cling with bleeding fingers to a bare rock face, struggle against impossible water odds, heave up immovable weights to free themselves'.

The reluctant aviatrix *does* get the plane down, with both herself and it more or less in one piece. She then even manages to overcome her fear of flying sufficiently to allow herself to be flown back to Dick at Brinalli Downs (all other routes are blocked by floods).

In her horrific ordeal Janet has learned not only something about piloting a plane but also a lot about herself. She's made up her mind, too, that it really *is* Dick whom she wants to marry (she had been temporarily deflected by an infatuation for his brother, Pete), and that 'People in real trouble are never as much alone as they feared they would be.'

Comfort in that thought is shared 'in dozens of home-steads scattered over hundreds of miles of the loneliest country in the world' as the click of the receiver going down for the last time at the Flying Doctor base signals Janet's successful landing. The 'weary pride' which the farmers have for their sometimes beautiful but always challenging country is a keynote of the book.

In *Adventure Stories for Girls* (1975) Hamlyn produced a typical example of the mercy mission for younger readers. Else Moorhead's short story, 'Daughter of the Flying Doctor', shows Julie Wayne helping her Australian Flying Doctor father in a clerical capacity, and by dealing with some of his radio calls. Like all full-blooded adolescent heroines, how-ever, she longs for a more active role: 'Gerry, I do wish you'd let me have just one little go at piloting the chopper!' Gerry Morgan, Dr Wayne's regular helicopter pilot, is adamant in refusing to agree to Julie's request, though he knows that Julie, who frequently travels with him through the outback to deliver medicines to remote homesteads, has learned a lot about flying simply by observation.

Julie's moment comes when she and Gerry are scouting for a villain who has stolen a stuffed kookaburra from Red-stones, an isolated bungalow owned by teenage Sally Howells and her brother Ted. (Julie has brilliantly and correctly deduced that the stuffed bird contains a fortune in gold sovereigns, intended for Sally and Ted.) Gerry has to climb down the helicopter's rope-ladder, clamber into a boat which the thief is frantically rowing away, and knock him out. Meanwhile someone has to keep the helicopter in the air and on course: 'Julie, do you think you can handle the chopper?' Gerry asks – and Julie, nerves tingling, stuttering incoher-ently, gasping and gulping, does so: 'Hardly daring to breathe, she moved the stick. And then she could have shouted for joy! The chopper answered, and dropped slowly, gently, lower into the chasm.' It is a case of 'Jolly good, Julie!' all round. Too bad that we shall never know if she ever flew professionally in Nancy Bird's footsteps, or whether her handling of the helicopter was to be a one-off feat.

Just as in the 1930s Nancy seemed to be symbolic of the Australian flying scene, Beryl Markham's name was uniquely linked in many people's minds with Africa, another impenetrable continent. Her place in aviation history rests largely on her becoming the first woman to fly solo over the Atlantic from East to West in 1936, but any assessment of her life tends to become a celebration of Africa, which she knew and loved so well, and at so many levels. As she said in her autobiographical *West with the Night* (1942):

> There are many Africas . . . Africa is mystic; it is wild; it is a sweltering inferno; it is a photographer's paradise, a hunter's Valhalla, and an escapist's Utopia. It is what you will, and it withstands interpretations. It is the last vestige of a dead world or the candle of a shiny new one.

To Beryl it was also home. Although she was born at Leicester in England, she went to East Africa at the age of four with her divorced, ex-army captain father, Charles Clutterbuck, who created a farm there from a stretch of wilderness. Beryl grew up playing with the local children and hunting with the Murani chieftain, Arab Maina; Swahili, Nandi and Masai seemed to come as naturally to her as English. Her formal education was minimal. She was expelled from several schools and made short work of a string of governesses. When her mother visited her in Africa, Beryl found her conventional, upper-crust English values 'boring'. Later on she helped her father with the breeding and training of racehorses.

In 1919, after a terrible drought resulted in the loss of his farm and fortune, Charles Clutterbuck went to Peru, but Beryl remained to become the first woman in Africa to receive a racehorse-trainer's licence. She was eventually to become one of Kenya's most successful trainers, with her horses winning six Kenya Derbys and many other prestigious events.

Beryl married, and fairly soon divorced, Jock Purves, a Scots rugby player. In 1927 she became the wife of Mansfield Markham who, after a spell in the Foreign Office, had, like

Beryl, become a horse-breeder in Kenya. The early days of the marriage brought Beryl into a very different environment from her African one. There was a Paris honeymoon, with clothes from Europe's top couturiers, plus a London season which caused speculation about her relationship with the then Duke of Gloucester. Mansfield Markham threatened to (but didn't) involve the Duke in divorce proceedings; the Markham marriage broke up and a generous sum from illustrious sources was settled on Beryl in 1929. Despite all this glitzy excitement she was later to say: 'I was unable to discuss the boredom of being alive with any intelligence until I had gone to London and lived there a year.'

There is no doubt that her capabilities were many and varied. She did not, however, shine as a mother. Returning to Kenya only a year or so after her son, Gervase, was born in 1929, she left him in the care of Markham's mother in England. Her African base then became the notorious Muthaiga Country Club near Nairobi, where many white settlers apparently engaged in adulterous and alcoholic exploits.

It has often been said that Beryl resembled Greta Garbo, and photographs of the finely featured, blue-eyed, blonde, arrogantly confident horse-breeder-cum-flyer suggest that this was true. The Muthaiga Club provided her with a range of willing admirers and devotees, but it seems that Beryl was often less interested in them as male animals than for the adventurous openings which they could provide for her. The American writer Martha Gellhorn aptly categorised Beryl as 'Circe casting a spell on Ulysses so that she could go along on his journey'. Apparently she 'knew what she wanted', and that was 'knowledge and adventure'.

Beryl's interest in flying began in 1931 when her friend Tom Campbell Black taught her to fly in a DH Gipsy Moth. He had come to Kenya from London to carry mail across Africa for Wilson Airways. He conveyed to his pupil an understanding not only of how to manipulate the machine, but also of the general difficulties of flying over Africa's vast and dangerous terrains. Beryl described Tom Black as both a careful and a casual pilot, who had previously 'never taught another soul to fly' and instructed her largely by 'intuition and instinct'. On several occasions he allowed her to get into difficulties in the air, taking over the controls only at the last minute when he felt sure that the lesson of that particular hazard had struck home: 'It's no good my telling you where you go wrong each time you do; your own intelligence will tell you that.' All this, with his dismissal of gadgetry except for trust in the compass, was to stand Beryl in good stead during her most demanding flights. She was full of admiration for Tom, who pioneered new routes for the mail, 'to probe inland Africa, seeking footholds for the future'. News of his death was eventually to come dramatically and poignantly to mar Beryl's enjoyment of the triumph of her Atlantic crossing in September 1936.

Denys Finch Hatton was another aviator (better known for his hunting expeditions) who was associated with Beryl. Her relationship with him disrupted his affair with Karen Blixen (who was, of course, the famous Danish novelist, Isak Dinesen). It was he who suggested to Beryl the idea of elephant-scouting from the air. Tom Black dissuaded her from going with Finch Hatton on his first aerial safari reconnaissance, and his hunch that something would go wrong seemed to be proved correct when the Gipsy Moth carrying Denys and a Kikuyu boy crashed, and both were killed.

Beryl obtained her commercial licence eighteen months after learning to fly. She became the first professional female pilot in Africa, flying mail and medicines and making good use of Denys's idea of hunting by air. (There seems to be some dichotomy in her feelings for Africa's animals; on the one hand her respect for wild life was tremendous, and often lyrically expressed in *West with the Night*; on the other, she could enter into the spirit of big game hunting. Nevertheless she was eventually to concede that 'It is absurd for a man to kill an elephant.')

Another of the Muthaiga Country Club set to become closely associated with Beryl was Baron Bror Blixen ('Blix'), who had, of course, formerly been married to Karen Blixen. Beryl had flown solo from Nairobi to London several times before she ferried Blix on this journey in a Leopard Moth in

March 1936. Their exploits *en route* read like a Hollywood film adventure. In Cairo, for example, they discovered a headless corpse, abandoned in a busy street. They were held up there for several days before the Italians would give them permission to cross Libya, which they occupied. In Benghazi – 'the fringe that harbours the useless ones of twenty nations, the cast-offs, the slag fallen to the side and forgotten until, out of necessity, it must sometimes be waded through' – the only accommodation they could find was in a brothel, and a dirty one at that. In Sardinia, Fascist officials held them for two days, suspecting both of spying and Beryl of being a man in disguise.

In London Beryl made up her mind to attempt to become the first pilot – woman or man – to fly from England to America. (Jim Mollison had previously made the east to west transatlantic trip, but he had started from Ireland, thus covering a shorter distance.) John Evans-Freke, 10th Baron Carbery, an English aristocrat who had abandoned his title and was a member of the Muthaiga set, agreed to sponsor Beryl's flight, although his comment on it – 'It's a deal . . . but gee, I wouldn't tackle it for a million. Think of all that black water! Think how cold it is!' – was hardly encouraging.

West with the Night records Beryl's anxious daily visits over three months to Gravesend, to see her new turquoise and silver Percival Vega Gull being assembled. Eventually she took off from Abingdon on her 3600-mile journey on 4 September 1936, with a just-about-acceptable weather report from the Air Ministry and a sprig of heather from her ground mechanic. More importantly, she had Jim Mollison's watch which he had loaned her as a talisman of her safe return: 'This is not a gift. I wouldn't part with it for anything. It got me across the North Atlantic and the South Atlantic too. Don't lose it – and, for God's sake, don't get it wet. Salt water would ruin the works.' Despite the fact that Beryl, failing to reach Yew York, landed in 'a nameless' Cape Breton swamp, Jim's watch remained undamaged. Beryl had accomplished a flight which made history, even though the behaviour of her fuel tanks was erratic and her engine had cut out, spluttered and restarted several times.

The flight provided her with celebrity, but not with wealth. In 1942 she made a third marriage, to Raoul Schumacher, a ghost writer who, it has been suggested, had a hand in the compiling of Beryl's moving and vivid *West with the Night*. This selective, impressionistic and extremely quotable biographical memoir omits a great deal (about sexual intrigues, marriage and motherhood, for example) but concentrates on Beryl's responses to Africa and its people, and her eulogies on flying and flyers. Ernest Hemingway's categorisation of it as 'a bloody wonderful book' when it was published in 1942 might have been somewhat ironical. Surprised that she could write so eloquently, he indicated off the record that 'some very fascinating' but apparently unflattering 'stuff' that he knew about Beryl had been carefully left out. Be that as it may, *West with the Night* provides a unique insight into the mind of a woman who threw off many conventional restrictions and took grandly and gorgeously to the air. It is also amazingly accurate in its prophecies about the future of aviation.

Beryl eventually left Raoul Schumacher in California, reverted to the name of Markham and returned to Kenya. Her successes there as a racehorse-trainer were resumed until in 1963 a virus attacked her horses and forced her to close her stables. She died at the age of eighty-three in 1986, 'poor and vodka-sodden', according to her current publishers (Penguin). However, she apparently managed to maintain to the end an aura of slightly tarnished glamour.

Her brand of iconoclastic independence hardly made her a role model for 1930s' writers of fiction for women and girls. However, the idea of girl aviators penetrating deeper into darkest Africa than anyone had done before offered distinct fictional possibilities. Ever since the early 1920s, authors in various weeklies had sent groups of schoolgirls off for the holidays to Africa, where strange or savage tribes had kidnapped them and threatened death, or worshipped them as Goddess Queens. There were also, of course, encounters with wild animals and discoveries of secret cities, hidden treasures, gold- and diamond-mines – and long-lost white heiresses. Dollops of diluted Rider Haggard still continue to crop up in girls' papers today, but during the 1930s the exploits of teenage aviatrices injected fresh vigour into these.

A typical example of wings-over-Africa fiction is the saga of *The Flying Sisters* by Ida Melbourne. ('Ida Melbourne' was actually Eric Lythe Rosman, a prolific writer who, under a variety of pseudonyms, wrote both dramatic and anarchically humorous stories for girls and boys. He had a special interest in aviation and during the Second World War volunteered for the RAF, serving as an officer in Bomber Command.) *The Flying Sisters* was originally serialised in the weekly *Schoolgirl* from no. 224 (11 November 1933) to no. 239 (24 February 1934) and published two years later in its entirety in *The Schoolgirl's Own Library*. It is a gripping and essentially colourful story in which the heroines Kit and Joan Fortune (aged twenty-one and fourteen respectively) have to tackle baddies (who are after the plans of the gold-mine which will one day make the Fortune family 'fabulously rich'), wounded lionesses, hostile giant apes, mischievous and highly organised bands of chimpanzees and two warring tribes of pygmies. As one of the characters eventually comments, all this suggests 'a travelling circus': the plot never gets entirely out of hand, but the author frequently seems to place his tongue firmly in his cheek. Nevertheless, despite the comic relief of overblown characters and situations, the aerial action is generally exciting and suggestive of authenticity.

Because the story was published in weekly parts, it is episodic and punctuated with cliff-hanging situations. It is also extremely complicated, with first one group and then another gaining the upper hand. The only consistent elements are the girlish grit and guile of Kit and Joan. Orphaned, with their elder brother Bob now 'the head of the family', they live in a bungalow in an undefined area of Africa. The story begins with Kit's plane apparently missing during her attempt to set a new record flying from the Cape to England: 'Kit, daring pilot, intrepid as any wartime aviator, had actually come to grief.' (She hasn't. She has just started late and gone slightly off course.)

Bob prepares to take off in an aerial search for Kit, and patronisingly refuses to take Joan with him – 'One flying sister is quite enough!' – but, because he is weaker and dimmer than the girls, he falls into a trap set by the villains and is abducted by them in his own aeroplane. So the sisters have to set off in Kit's plane, the *Sky Queen* (judging from the illustrations, this was probably an Avro 504), to find and rescue him. Joan, for whom 'the sound of aero engines roaring is music . . . as a bugle to a war horse', is at last in her skyways element.

Rather more cautious than Kit, she turns out in the end to be an even better pilot than her sister, whose harped-upon recklessness actually seems far removed from the clear, controlled and responsible attitude of most real-life female flyers:

> Kit, at the last moment, had flattened the machine, had hurled it forward at full throttle to climb steeply upwards. A mighty tree, that had stood the test of hundreds of years, seemed to ride before them, immense, menacing . . . Joan shot a side-glance at her sister's face . . . she was actually humming a song. In her eyes was the light of battle and joy, in her cheek a spot of colour.
>
> 'Goodness!' was Joan's thought. 'She loves this – she loves the risk!'

They miss the tree 'by a mere foot'; Joan laughs 'jerkily', compliments her sister on her skill, but asks if it were necessary to go so close. Kit, with eyes 'twinkling and dancing', simply replies, 'Was that close?'

There are lots of exhilarating flying sequences, including a 150-mph dive: 'The screeching of the wind in the bracing wires, the whine of the propeller, were deafening . . . "Phew! Was I near death, or was I not?" Kit asked herself.' In the air, on the ground, in lions' lairs, pygmy prisons, incredibly dense forests and arid wastes, the girls 'brave all for the sake of their brother in peril', and defeat strong and wicked men by sheer brain-power: 'Men may be strong, and they may have revolvers, but it isn't that that counts; it's wits.' And, of course, they do find and free their brother, as well as rescuing from captivity another English male, Chalmers, who had been Bob's partner when the gold-mine was originally discovered.

After all this, the physically injured and psychologically rather useless Bob has the nerve to suggest, when they are rushing to register the gold-mine claim, that Chalmers – and not the girls – should pilot the plane. As one would expect, he is fiercely put in his place.

It is not exactly Haggard, Hemingway – or Markham – but aspiring aviatrices or explorers amongst the *Schoolgirl* and *Schoolgirl's Own Library* readership might well have taken inspiration from *The Flying Sisters*, particularly from fourteen-year-old Joan, who, untrained apart from reading flying manuals and observing Kit at the controls, undergoes her baptism of piloting fire when her sister faints in the air after being wounded in a fight with 'a savage'. From then on Joan acquits herself so well that, on her eventual return to civilisation, she is 'by special permission given a licence although under age'.

Contemporary cartoon showing the early British balloonist, Mrs Graham, falling from a balloon but saved from catastrophe by her billowing skirts.

1.

A glamorous and unrealistic interpretation in the *Illustrated Police Budget* of Dolly Shepherd's mid-air rescue of Louie May.

2.

3. Dolly Shepherd in her bloomered
    parachuting attire.

4. Harriet Quimby, the first woman to
    pilot a plane across the English
    Channel.

5. First World War Women's Royal Air Force recruits.

6. Katherine Stinson arriving in China in 1917.

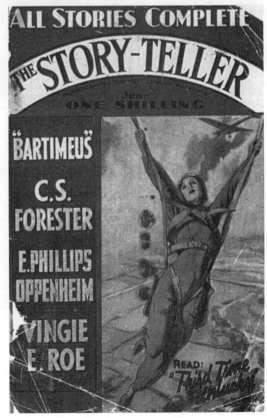

7.

8. First World War driver attached to the Royal Flying Corps
(from *The Sphere*, 1918).

Left: *'The Duchess of Bedford is a fearless airwoman who has flown to India and back.'* Below: *'Here we see Lady Bailey, who is another courageous supporter of flying for women'.*

(From *School-Days*, 1930.)

Bottom left: Linda Rhodes-Moorhouse and her son, William.

Bottom right: Mary Tourtel – aviation enthusiast, and the creator of Rupert Bear.

10.

11.

12.

LITTLE MISS AEROPLANE. A Complete Story about a Plucky Little Girl.

13.

(From the comic *The Sunbeam,* 1922.)

Left: Amelia Earhart in 1928.

14.

Flyin' Jenny, the picture-strip heroine
in the Amelia Earhart mould.

15.

'Keeping an aeroplane of
one's own.'
(From *Woman's Magazine*,
1930.)

**I wish I were
an angel
mummy-**

I wish I were an angel, mummy,
I would so love to fly.

Hold your tongue, precocious child.
You may be by and by.

But can't I do like daddy did
To " gratify his whim—

Learn at 40 bob an hour
In aeroplanes at Lympne "?

I heard him telling Uncle John
" It's grand up in the sky."

Oh, must I wait until I'm dead
Before I learn to fly?

16.

GET YOUR TUITION IN THE HERE-BELOW AT

LYMPNE

**CINQUE PORTS
FLYING CLUB** **KENT**

Expert ★
Instruction

Repairs ★
and Service

Sales and ★
Stock

HYTHE 1617

(From *Popular Flying*, 1937.)

£10,000 To Be Won In Eight Weeks
First Easy Pictures To-Day

and Town Talk

Ideas

2D

VOL. LVIII—No. 1507   JANUARY 27, 1934

Registered for Transmission
by Canadian Magazine Post.

"PLANE JANE"

ARTHUR FERRIER

17.

18. Jean Batten acknowledging her welcome at Sydney Airport in the 1930s.

19. Jean Batten in her seventies.

20.

21. Beryl Markham, from the cover of
    The Illustrated London News,
    2 September 1936.

22. Nancy Bird – with the cushions she
    used to raise her height in the
    pilot's seat.

23.

ONE man offered her soft lights and sweet music. The other, a ring, a home, and no nonsense. But it's not so easy as all that for a girl to make up her mind, is it?

"*I* KNOW— *I Married Him*"

24.

(From *Home Companion*, 1942.)

MAY'S eyes were fixed wistfully upon the leader of the band.

Pauline Gower in ATA uniform   25.
(From *Woman's Journal*, 1942.)

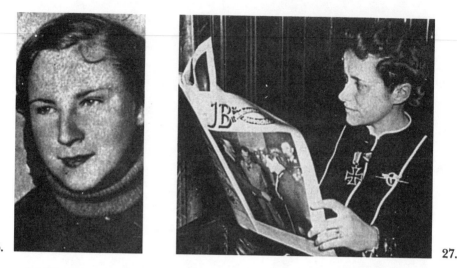

26. 27.

Left: Lieutenant Lily Litvak, the White Rose of Stalingrad. *Right:* Hanna Reitsch reading an account of herself receiving the Iron Cross.

28.

29. Jacqueline Cochran and Jacqueline Auriol, constant competitors for speed records during the 1950s.

30. An American stewardess from Frontier Airlines checks that her grooming is up to standard.

31. (From *Girl*, 1958.)

32. *Left:* Professor Peabody and Dan Dare in 1950s *Eagle*.

Below: Professor Peabody and Dan Dare from *New Eagle*, 1990.

33.

34. (From *Wonderwoman*, 1976.)

35. The Stargrazers – 'Sexy sailors of the Spaceways', 1989.

36. Sheila Scott.

37. Jeanna Yeager.

38.
Valentina Tereshkova from the USSR, the first
female astronaut.

39. Helen Sharman, the first
British woman to go into
space, in 1991.

40. Hilda Hewlett, the first British woman to acquire a pilot's certificate, in 1911.

41. Flight Lieutenant Julie Gibson, who acquired her RAF wings in 1991.

43. Flying Officer Anne-Marie Dawe, Navigator with the Fleet Air Arm, 1991.

42. Dolly Shepherd in old age with parachutists of the RAF Falcons.

Ellen S. Baker

Mary L. Cleave

N. Jan Davis

Bonnie J. Dunbar

Anna L. Fisher

Linda M. Godwin

Marsha S. Ivins

Mae C. Jemison

Tamara E. Jernigan

44. A selection of NASA women astronauts.

Shannon W. Lucid

Christa C. McAuliffe

Chiaki Mukai

Ellen Ochoa

Judith A. Resnik

Sally K. Ride

M. Rhea Seddon

Kathryn D. Sullivan

Kathryn C. Thornton

CHAPTER NINE

Winning Through with Worrals — and the ATA Girls

WORRALS FLIES AGAIN

by Captain W.E. JOHNS

Author of The "Biggles" books

HODDER & STOUGHTON
LIMITED   LONDON

'Four or five times a week for five months I've been taking battered Tiger-Moths back to the makers for reconditioning. It's about as exciting as pedalling a push-bike along an arterial road.'

CAPTAIN W.E. JOHNS
Worrals of the WAAF (1941)

THROUGHOUT the 1930s the *Girl's Own Paper*, issued by the Religious Tract Society (now functioning as the Lutterworth Press), endeavoured to echo the popular mood and become airborne. As well as publishing flying stories by Dorothy Carter and others (see chapter 7), it gave away art plates of celebrated aviatrices, ran articles on careers in aviation and held regular competitions through its Skylarks Club ('Write an essay of not more than 500 words on "Why I want to Fly"'; 'Write a letter to your MP demanding an Air Training Corps for Girls'; 'Make Your Own Aeroplane!', etc.).

However, because the *GOP* was proud of its internationalism and its large circulation abroad, it ignored the disturbed political atmosphere which was intensifying in Europe during the last years of the decade. A world-wide club flourished through the paper; members had the satisfaction of seeing their photographs reproduced in it, and of corresponding with each other. Stephen Spender has suggested in his book *Citizens In War – And After* that 'The Vicarage and the Women's Institute did not wake up to Fascism until the time of Munich.' This, of course, was in 1938, when awareness of the real meaning of Fascism was growing at home through the rhetoric and rallies of Mosley's British Union of Fascists, and the influx of Jewish refugees escaping from persecution in Nazi Germany. The *GOP* could hardly reflect all this, and it valiantly tried to see the best in Germans and Germany with an enthusiastic article in 1937 about the girls of the Hitler Youth organisation. Some of the Gerdas and Grizelas who were described as selflessly helping the Third Reich on its way by 'practising handicrafts, taking courses in physical education and household management, and spending much of their leisure time in making clothes for needy families' were doubtless *GOP* readers. They, just as much as their English fellow Club members, would have appreciated the paper's photographs of earnest teenagers working for their 'A' licences, Dorothy Carter's March 1938 article, 'Flying as a Career for Girls' (which included a picture of the young German test pilot, Hanna Reitsch), and, in early 1939, uncontroversial flying stories such as 'Kites, Worries and Ink' by Pauline Gower (who was soon to be heading the wartime women's section of the ATA) and 'Five Bob Flights' by Ruby Evans.

Even in the early days of the war the *GOP* gave it low-key treatment, suggesting that girls should concentrate as much as possible on their knitting, school-work and everyday routines. It was not until the paper's October 1940 issue (obviously planned and put together some weeks before this) that, shattered by the fall of France and the Dunkirk evacuation, and inspired by the Olympian achievements of 'the Few' during the Battle of Britain, the editorial tone changed dramatically. The struggle against Hitler and all he stood for was now being taken seriously: declaring, 'We've all been shaken out of our ruts and we have all got to show what we are made of today as never before,' the editor triumphantly announced,

> Behold Worrals – she's emerged to do her bit in this war. She's a real live character with a genuine love for planes; in fact her job means something to her. It is her whole life. She's ready to take all personal risks in the service of her country. She's made up her mind that there'll always be an England.

Stirring stuff – and there was little doubt that the new heroine would live up to expectations. She was after all the brain-child of Captain W. E. Johns, whose stories of Biggles were lapped up avidly not only by boys but by girls as well.

With his RFC and RAF background, Johns was a commendable amalgam of orthodoxly accepted values and aeronautical panache in the *GOP* context. Pilot Officer Joan Worralson ('Worrals') appeared on the fictional scene at exactly the right psychological moment to epitomise the bloody-mindedness of the British with their backs well and truly against the wall. Involved with RAF recruiting during the Second World War, Johns apparently created her in response to an Air Ministry request, because girls were not dashing to don Air Force Blue at the same speed with which they were entering the WRNS or the ATS. The efficiency of Worrals as a stimulus to recruitment is indicated by the fact that, only a few weeks after the first serial about her began, the WAAF had temporarily to halt its intake of new volunteers. (Of course, the Battle of Britain's polishing of the RAF's *Per Ardua ad Astra* image might also have had something to do with this.)

It is interesting to note some discrepancies and mistakes in this first Worrals story which were surprising, even with allowance for 'poetic licence', from an author who was so knowledgeable about the air services. In the *GOP* Worrals is designated as a Pilot Officer, but no such rank existed in the WAAF, although it was introduced after the war when a regular service, the Women's Royal Air Force (WRAF) replaced the wartime auxiliary one. When the story was published later on, towards the end of 1941, in book form, Worrals's rank had tactfully and authentically been changed to Flight Officer. In both the serial and the book, however, Worrals is incorrectly described as wearing 'the coveted "wings" on the left breast of her tunic'. Waafs were not permitted to wear the wings brevet nor to fly (female flying jobs were the province of the ATA).

Before he wrote *Worrals of the WAAF* Johns had for some time been supportive of women in aviation. He had devoted the June 1934 editorial of *Popular Flying* (the monthly which he edited for many years) to this theme, under the heading 'Women and Wings'. 'They are going to fly,' he told his adult readers firmly, 'make no mistake about it, the average girl flies as well as the average man!' He saw women not only as civil pilots but as potential military ones, 'lady aces in a national emergency'.

By October 1940, when the story began its serialisation in the *GOP*, Captain Johns seemed to welcome the challenge of originating a female character. By then the Biggles saga had possibly become a bit of a sinecure; at any rate, he put some of his best writing into Worrals' early wartime adventures, and conveyed considerable empathy with the aspirations of young girls. Worrals and her side-kick, the blonde and fetching Betty Lovell ('Frecks'), colourfully personified not only the patriotic fervour but also the frustrations which the war aroused in so many female adolescents. Bursting to do their bit, they were often allowed only to sew shirts or knit socks for soldiers and refugees. Some were allowed to help the Red Cross, to collect salvage and to grow vegetables on allotments at home or at school. Few felt that they were contributing to the war effort as vigorously as they would like to have done.

It is interesting that of the few cult figures who emerged in the juvenile papers of the Second World War, Worrals – a girl – is the most memorable. As Graham Greene wrote in his essay 'The Unknown War' in 1940, 'Perhaps the spirit of these heroes is best exemplified by a heroine – Worrals . . . we hear of her first as she sat moody and bored on an empty oil drum, complaining of the monotony of life.' He is referring to the opening scene of the story, when Worrals is suffering the same restrictions as the skilled women pilots who in their early days with the ATA were allowed to ferry only a limited selection of planes. Some advanced models were reserved for male aviators. Strictly speaking, of course, Worrals shouldn't have been ferrying planes at all, because Waafs never piloted aircraft.

At least one real-life Waaf did, however – involuntarily – go up in a Spitfire. At Hibaldstow airfield in 1945, A.C.W. Margaret Horton, a wartime air-frame mechanic, was detailed to sit on the tail of the Spitfire AB910, to hold it down during the engine run-up. Her presence was unknown to the pilot, who took off and completed a ten-minute flight over the airfield with Margaret clinging to her perch on the tail of the machine.

One wonders whether girls inspired by Worrals to join the WAAF found life there rather tame after the gloss of their heroine's flying exploits. Only eighteen years old, she eased throttles, roared engines and sped her machines through the battle-scarred skies as niftily as any much-experienced male fighter ace. Her confidence in such hazardous situations spilled over into confrontations with Nazi spies and secret police, the Japanese, Middle Eastern gun-runners, dope-pedlars and slave-dealers. Astoundingly, despite her dealings with so many baddies, Worrals manages without a great deal of effort to retain her virginal state. It is also strange, given the wartime settings of the early stories, that two such appealingly nubile girls as Worrals and Frecks, so often surrounded by servicemen, were not snapped up as wives or sexual partners. But one has to bear in mind that Johns was consciously writing for twelve-year-olds as well as for those who might be approaching the age when they could join the services. Underplayed epithets for man–woman relationships occurred in his factual articles for the *GOP* as well as in his fiction. For example, in his regular 'Between You and Me and the Joystick' feature, in July 1941, he points out that WAAF recruits have 'lots of fun' in off-duty times, 'one pastime being roller-skating in the hangars. Incidentally there are no regulations against romance and, from what we hear, the people who sell engagement rings [a euphemism here, perhaps?] near aerodromes are doing a brisk business.'

Naturally some men *are* rash enough to make passes at the dynamic airwoman duo. There is a chilling moment in *Worrals Carries On* when, operating clandestinely in Nazi-occupied France, they are 'ogled' by members of the Wehrmacht. Generally speaking, however, men sense that they will receive short shrift and do not push their luck. But Worrals does have one long-standing admirer. He is the handsome Spitfire pilot Bill Ashton, who fancies her from the first chapters of the saga. There are implications that she has a soft spot for him too (allowing him the physical fulfilment of helping her on with her mackintosh), but when things threaten to get too passionate, and Bill desperately declares, 'You know, kid, you mean an awful lot to me. If anything happened to you, I should never forgive myself,' Worrals firmly puts him down: 'Be yourself. You'll laugh at this nonsense in the morning.'

The stories are spiced with Worrals' own chauvinistic brand of feminism: 'To call a man by his first name is to put all sorts of conceited notions into his head.' Worrals' only lasting relationship is with Frecks, and their friendship is an inspiring one. Frecks has a flippant approach, but she is as gutsy as Worrals when it comes to a crisis. Her fluffy blonde-ness is a perfect counterbalance to Worrals' dark-haired looks and more serious nature. Their friendship is convinc-ingly portrayed and deepens as the series of books (eleven in all, several of which were serialised in the *GOP*) progresses, so that, when the war ends, it seems natural for them to live together and to share several civilian air-detection enterprises.

A recurring theme in the books is the necessity to kill. When Worrals (unlike any real-life Waaf or ATA pilot) has to shoot down a German monoplane in the first story, she is portrayed with appropriately sensitive responses – dry lips, cold hands and moral doubts – but she presses the machine-gun button nevertheless: 'I've got to do it – this is war.' Although he was writing for girls, whose work in the services and on the home front would normally be non-combative, Johns realised that, if the stories were to carry any clout in the climate of wartime, occasional killings by his heroine would have to take place. As the saga (and the actualities of the war) proceeded, Worrals became less hesitant about assuming a militant role: 'Frankly, I'm only concerned with getting home, and if a few Huns get hurt on the way I shan't saturate my pillow with tears on their account' (*Worrals Carries On*, 1942). By 1949, when in *Worrals in the Wastelands* she is hunting for Anna Schultz, the Nazi 'she-devil' of an internment camp (obviously based on real-life Irma Grese, 'the blonde beast of Belsen'), she comments, 'I'll hang her for you if you like – and so would any other woman who knows her ghastly record.'

As well as stirring readers with the thrills and chills of Hitler's war, the Worrals tales put across more vividly than any pre-1939 flying stories for girls the virtues of courage, teamwork, loyalty to cause and country, and the value of friendship. The heroine has become a role-model for all this, and for her inspirational tackling of the technicalities of aviation: 'for the sheer joy of it Worrals roared through the solitude, the aircraft skimming the cloud tops like a dolphin in a silver sea . . . Then she drew the control column back and climbed steeply towards the sun.' Her aeronautical know-how is also put to good use when she is trapped in a plane with German spies who plan to fly her to the Fatherland 'where they know how to control impetuous young women'. She grabs the controls and flies so low that they dare not touch her, knowing that she would then suicidally crash the machine and kill them all. (She has nothing to lose, for the alternative is a German concentration camp.)

Marise Duncan, the heroine of Dorothy Carter's series of six books (see chapter 7), becomes, like Worrals, involved with the Gestapo when she has force-landed behind the German lines (*Sword of the Air*, 1941). She is ferrying fighters to France in the autumn of 1939 and has apparently antici-pated the formation of the women's section of the ATA by several months (Pauline Gower was not authorised to recruit the first eight female pilots until January 1940). Marise's adventures are rather over the top in this particular adven-ture, and coincidences abound. She meets up with her father who, as 'Herr Ebermann', is working for British Intelligence, and, as 'Magda Vorenstern', Marise also embarks on espio-nage as a worker in a Messerschmitt 110 factory. We are told that 'The Duncan family had always been good at languages', and certainly this must have been so, for the young and utterly untrained Marise to slip so surely into work that would have taxed the abilities of experienced girls in the Special Operations Executive.

Marise's Russian adventures are described in chapter 10, and, in the last book in the series, *Marise Flies South* (1944), she is flying bombers out to Australia and New Guinea and tangling, both in the air and on the ground, with the

Japanese. The saga ends on a hopeful, hands-across-the-sea, flyers-of-the-Allies-unite note. Marise and her American friend Lieutenant Pete Wilkie declare that 'As long as we [the English-speaking people] stand together we ought to be able to do something for this silly old world. And not in the war only, but in the peace afterwards.'

Adult fiction of the Second World War paid little attention to Waafs or female flyers, although the former were used in numerous attractive magazine advertisements to promote everything from toothpaste to Tangee Lipstick, from hair-nets to home furnishings and gas-cookers. These had nothing, of course, to do with flying but, presumably, a little of the RAF's Silver Wings lustre vicariously rubbed off on the girls in Air Force Blue and made them appealing as product promotors. One of the most engaging 'pin-ups' of the war was the smiling, brunette Waaf pictured on the cover of the November 1940 *GOP*. Painted by Manuel McKinlay, a Portuguese artist who had a flair for portraying healthy, wholesome yet sexy-looking girls, this picture was considered by thousands to be a representation of Worrals. In fact, as I realised only in the 1990s, it couldn't have been, because McKinlay's Waaf wears the badge of an aircraftwoman and not, as Worrals would have done, of an officer.

One Waaf who *does* get a showing in an intriguing work of adult fiction is Prudence Cathaway in *This Above All*, by Eric Knight (1941). She gets involved in a rather improbable and frustrating love affair with Clive Briggs, a soldier who be-comes a deserter after disillusionment with the military ethic when serving with the British Expeditionary Force in France. Prudence tries to persuade him to rejoin his unit, but a final decision is never made because he is killed when rescuing a woman and child from a blitzed building. The challenging discussions of the comparative merits of British and German societies do not deeply involve Prudence, although both she and Clive share a general sense of futility about the war. Prudence is upper-class, and her personal battle is with the snobbishness that forces her to think in terms of 'common private soldiers', 'workingmen's public houses', and so on. Clive, aggressively aware that he is from the people, pours salt into her sensitive, self-inflicted, psychological wounds. *This Above All* has none of the warmth of *Lassie Come Home*, the book for which Eric Knight is best remembered, although when it was filmed in 1942 Tyrone Power mellowed the acerbic character of Clive, and Joan Fontaine looked great in WAAF uniform.

On the whole, in adult stories written during the 1940s, Waafs could just as well have been Wrens, Ats, nurses or girls on factory assembly lines. They were simply backers-up of men, and girls to whom the soldiers, sailors and airmen would one day come home. *Home Companion* (30 May 1942) casts May, a clerk in the WAAF, in this role in Eleanor Harvey's story, 'I Know, I Married Him'. Soldier Micky is May's steady boyfriend, but she begins to make a fool of herself with Elmer Mace, a band-leader, 'forty if he is a day', who, unknown to her, is firmly married to someone else but very happy to play around. May dreams of orchids and champagne (particularly potent symbols in the austerities of wartime, one imagines), but finally, of course, gets the brush-off. Fortunately the stalwart Micky comes home on leave to pick up the pieces, and May is sufficiently recovered to return to her WAAF duties and participate in the war effort with new zeal and determination. The story's lead-in really tells all: 'One man offered her soft lights and sweet music. The other, a ring, a home, and no nonsense. But it's not so easy as all that for a girl to make up her mind, is it?'

As well as writing flying stories for girls in the *GOP* before and during the war, Pauline Gower eventually took over its factual aviation feature, when W. E. Johns gave up 'Between You and Me and the Joystick' after November 1941. She wrote several articles for adult magazines (mainly about the work of the women of the ATA) as well as the autobiographical book *Women with Wings* (1938), which tells the story of her partnership with Dorothy Spicer during the 1930s, first in an air taxi service and later a flying circus act.

Pauline Gower was one of the most interesting women in British aviation. She came from a privileged background and, as the daughter of Sir Robert Gower, MP, did not have to earn her living. As Pauline says in 'Women Who Fly Alone'

in *Woman's Journal*, April 1942, she 'went through a London
"season", was presented at Court, did all the things expected
of a débutante, and was bored to tears with it'. Her reason
for taking up aviation was an unusual one: 'having been
stricken by a pernicious ailment which barred me from all
active sports at the age of seventeen . . . flying appeared to
me the ideal "sedentary" occupation.' She ran away from a
finishing school in Paris and learned to fly at Stag Lane:
'Here I met Dorothy Spicer and Amy Johnson . . . We were
all three staunch friends.' They were all, of course, also
thoroughly professional and well qualified. Dorothy Spicer
became the only woman then to hold all four ground engineer's licences; Pauline, after getting her private and commercial licences, took the second-class navigator's certificate
and a GPO wireless operator's licence.

Her enthusiasm for her chosen profession, even at a time
when her ATA duties must have been heavy, spills over
infectiously: 'For me, aviation is top-ranking among all the
careers open to women. More than that, I would say that
every woman should learn to fly. Psychologically, it is the best
antidote to the manifold neuroses which beset modern
women.'

As well as being commandant of women ferry pilots in the
ATA, Pauline Gower became a board member of the British
Overseas Airways Corporation while the war was still on, and
was the first woman anywhere to serve thus on a state airline.
She continued to write 'The World on Wings' features for
the *GOP*, but the issue for May 1947, instead of including
one of her articles, contained an editorial obituary tribute to
her. It mentioned highlights in her flying career and the fact
that, in 1945, she had married Wing Commander William
Cusack Fahie. It sadly laments the sudden death of 'this
brave and clever woman' and adds: 'So keen was she on her
GOP work that her last aviation article for it was written the
day before she died.' What it doesn't say is that she died in
childbirth, bringing boy twins into the world – truly a tragic
way for a high-flyer to be grounded.

Pauline's partner Dorothy Spicer ran an aerial garage for a
year after the two girls terminated their partnership. She

married, and did not become an ATA pilot during the war because of her family commitments. Bizarrely, she too died soon after the ending of hostilities, when she and her husband were involved in an air crash in South America. It seemed unbelievable that the two talented, attractive and vibrant young women who had so insouciantly but effectively made their living from flying in the 1930s should both be dead just a decade afterwards.

Their ingenuity in making their venture pay was impressive. For much of their time together the two girls lived in primitive conditions in a gipsy caravan; they flour-bombed a speed-boat (by prior arrangement with its passengers) as a publicity stunt for their aerial circus, they arranged parachutist displays, crazy flying and, of course, the inevitable and popular joy-rides (starting at 3s 6d and rising to as much as 10s for longer trips).

In October 1938, when war in the fairly near future seemed a probability rather than a vague possibility, the Civil Air Guard (CAG) was formed. This was a scheme to encourage the training of new pilots, who would be desperately needed once hostilities began. Fifty-seven of Britain's fifty-nine flying clubs participated; subsidised flights became available at half a crown (instead of something like £2) per hour. In less than a year after the beginning of the scheme 10,000 new pilots were in training, nearly 1000 of whom were women. They were not all successful in obtaining pilot's licences, but several of those who did were able to make a valuable contribution to the war effort. Charles Grey Grey (always known as C.G.G.), the cynical and, one suspects, antifeminist editor of *The Aeroplane*, condemned the CAG because of female involvement. He did not mince his words: 'The menace is the woman who thinks that she ought to be flying a high-speed bomber, when she really has not the intelligence to scrub the floor of a hospital properly.' C.G.G., who had founded *The Aeroplane* in 1911, was its editor right up to 1939, and all the most outrageous and often absurd quotes from that otherwise excellent journal are from his pen. In the late 1930s he totally succumbed to Nazi blandishments – they invited him over on various occasions to see their latest aircraft – and his writings in 1938–9 are all about how wonderful Hitler's Germany is. Grey was not really a Nazi, but just liked all the VIP treatment he received. Anyway, the anti-feminist remarks of 1911 and of the 1930s are by the same man (see chapter 2).

By the time Pauline Gower joined the ATA, she had accumulated 2000 hours of flying experience. Altogether this ferrying organisation was to employ over 150 women pilots, and about three times as many men. Three hundred thousand different aircraft were to be safely delivered to airfields during the course of the war. The pilots came from differing backgrounds; some were affluent spare-time flyers, some professionals and some had been trained by the CAG. Despite opposition from several men in the ATA, Pauline was deputed in December 1939 to form a female ferry pool at Hatfield.

She took on the assignment with relish, appointing eight pilots—Winifred Crossley, Margaret Macdonald Cunnison, the Hon. Margaret Fairweather, Mona Friedlander, Joan Hughes, Gabrielle Patterson, Rosemary Rees and Marion Wilberforce. Their ranks were soon to be extended, but not at first as quickly as had been hoped. Pauline took the opportunity of asking the American aviatrix, Jacqueline Cochran, who wanted to get into the war, to find some 200 already trained female flyers from the United States who would come to serve on the ATA. After diligent research, Jacqueline came up with a much shorter list; nevertheless the twenty-four pilots whom she recruited proved an invaluable addition to the Hatfield ferry pool.

It was a bone of contention that women were at first allowed to ferry only 'the light type of trainer aircraft from factory to aerodrome', while there were no restrictions on those which the male pilots could fly. However, common sense soon prevailed, and it was acknowledged that competent women flyers could cope with anything from Tiger Moths to heavy four-engined bombers. (The only craft that remained vetoed for the ATA girls were flying-boats. This was not because of any supposed incapacity on the part of the pilots, but simply because ferrying them might entail overnight stops on board with crew-men. So the ban was in defence of the girls' honour.)

Essentially the ATA pilots had to be resourceful and able to adapt quickly to different machines. There was rarely opportunity to practise in advance on an unfamiliar aeroplane. The girls were simply, before take-off, given a sheaf of handling notes to be strapped to a knee for reference during the flight. Responses from the personnel on the receiving-ends of the planes varied from surprised pleasure to resentment when they discovered that the ferryers were female. Eventually, however, when the war effort cranked up at an astounding rate, and there was a truly desperate shortage of pilots, 'They didn't mind if you were a man, woman or a monkey', according to Lettice Curtis, an ATA member who has recorded her memories in a fascinating book called *The Forgotten Pilots* (1971). Other books of reminiscences of their wartime ferrying days have been produced by Ann Welch (*Happy to Fly*, 1983), Lady du Cros, *née* Rosemary Rees – (*ATA Girl*, 1983) and Alison King (*Golden Wings*, 1956). The writers have in common a wry sense of humour and an unpretentious approach to their important and hazardous wartime work. The planes they flew were unarmed, so they would have been in danger had they come within range of enemy aircraft. The main problem they faced, however, was our good old murkily unpredictable British weather.

Over 100 ATA pilots were killed in flying accidents during the war; fifteen of them were women. Although they often had to cope with prejudice and discrimination (even being paid £80 less per year than the men who did exactly the same job), there was a general sense of sadness when the ATA began to run down in the middle of 1945. It had taken the horrific background of war to provide them with the opportunity, but the ATA girls had found tremendous satisfaction in piloting aeroplanes which they might otherwise never have flown. Many of the aviatrices hoped to continue in some sort of flying career after the war ended, but few were successful, and ten years after their release from the ferrying service only seven of the female pilots were still flying professionally.

Many of the girls saw themselves then and later as the forgotten pilots, neither fully accepted by the RAF nor acknowledged by the public. It was not until 1988 that their efforts were celebrated in a best-selling work of fiction: *Till We Meet Again* by Judith Krantz. This is a several-generational story which sprawls over France, England and America, with all the glitz, glamour and sexual excitement that this author generally provides. Marie-Frédérique ('Freddy') wants to fly when she is only three years old. Growing up in America, she finds the boys and college men who can't talk intelligently about aviation boring. With her blue 'saturated with sky' eyes and 'bonfire' hair, she looks good enough to become a movie star like her sister, Delphine, but flying comes before anything with her. She stunt-doubles in Hollywood for Alice Faye, Constance Bennett and Nancy Kelly in *Tail Spin*, wins several small race events, and gets good places in big ones. She encounters prejudice against women instructors and is incensed to realise that any far less experienced male inspires more confidence in pupils than she does: 'Did they think that you flew with your cock?' she asked indignantly.

Poor little war-torn England seems to offer greater career satisfaction than America. The CAG has just been formed, so Freddy joins it as an instructor; soon after the outbreak of war she becomes one of the ATA's women pilots. Amy Johnson and several of the other real-life flyers come into the story as part of 'the most splendid and honourable company of women pilots to be found anywhere in the world'.

Freddy has the normal eye of an attractive girl for men, and marries Tony, the squadron leader brother of one of her ATA colleagues. Her excesses of passion, however, seem to be reserved for aeroplanes, as when she falls in love with a Mark 5 Spitfire:

'I love you, I love you,' thought Freddy in rapture. 'I love every one of your one thousand two hundred and fifty fierce and mighty horses, I love the clear bubble of your perspex canopy, I love your tapered, ellipse-shaped wings and your noisy, don't-give-a-damn exhaust and your snug narrow cockpit and your crazily crowded instrument panel . . .

(The eulogy continues for over half a page.)

Freddy eventually returns to America, and feels further rapture at free-flying again in wide open spaces after the constricted ATA routes in England. Her romances are threaded between her flying fulfilments, and *Till We Meet Again* – especially the section set in the world of the ATA – is satisfyingly atmospheric and exciting. I loved the plot, adored the flying scenes, and introduced the book to a pilot friend. She enjoyed it too, but commented that Judith Krantz's concept of 'what the rudder is for' is somewhat hazy. Nevertheless this celebrated author *did* give the ATA girls the credit which they so much deserve.

As a footnote to the experiences of wartime aviatrices, it should be recorded that women can still fly Spitfires, as was shown in September 1991 at an air display at White Waltham before an audience of mainly women pilots. Carolyn Grace (born Carolyn Mansfield on a stud farm in New South Wales) expertly demonstrated the manoeuvrability of the two-seat Spitfire IX restored by her late husband Nick Grace. When he was killed in a car crash in 1988, Carolyn decided that the best memorial to him was to keep his Spitfire flying. She accordingly took lessons from an airline pilot, Peter Kinsey, and mastered the art of flying this elegant thoroughbred aeroplane.

CHAPTER TEN

Iron Crosses and Russian Roses

'Let's get the fighting over first, darling – then maybe we can talk about love.'

LIEUTENANT LILY LITVAK,
the White Rose of Stalingrad

DURING the Second World War, in common with the British and American armed forces, those of Nazi Germany refused to accept women pilots for combat duties. However, one of their female flyers, Hanna Reitsch, achieved fame – or notoriety – as 'Hitler's test pilot'.

A somewhat austere and controversial character, Hanna's enduring passions were flying and love for her country. It therefore seems ironic that although she was able to use her outstanding flying skills in the service of the Fatherland at a time when they were most needed, acclaim from many of her compatriots was relatively short-lived. Unlike male aviators of the time, she was apparently victimised after the war simply for having carried out her normal duties. The question of whether or not she was sympathetic with Nazi ideology was constantly raised after the time of the Allied victory, although motives of other flyers and of servicemen were not similarly questioned.

Religion played an important part in Hanna's childhood, and, even when she was a young woman and a celebrated pilot, her mother sent her regular admonitions about the dangers of pride and the need to place herself in God's hands. In her early teens, with an interest in medicine possibly inherited from her opthalmic specialist father, Hanna developed an ambition to become a flying medical missionary. In the end, the challenges and satisfactions that she found in aviation pushed aside her medical aspirations. Born in 1912, she grew up at a time when, under the restrictions of the post-Great War Treaty of Versailles, powered flight and the rebuilding of an air force were taboo in Germany. In the early 1920s, enthusiasts circumvented this ban to an extent, by taking up gliding in a variety of converted planes or improvised engine-less aircraft.

Like so many would-be aviatrices, Hanna, who was only 5 feet tall and extremely slight, found that few people took her ambitions seriously. She was the only girl on one of her first gliding courses, but soon had the distinction of breaking a couple of records. The first, for endurance, was intentional, when she stayed aloft for five hours; the second, for altitude, was involuntary, and was achieved when she was pulled up through a storm cloud far higher than she intended to go. She admitted to fear on this occasion, when ice froze the glider's instruments and, losing control, she was battered by hail and wind.

Her skill and determination soon elicited appreciation, and she was invited to join the Glider Research Institute at Darmstadt. She became internationally respected as a glider pilot and represented Germany at events in European countries and America. She was also one of a group of German glider pilots who successfully flew over the Alps. Her first assignment as a test pilot was for the Darmstadt Institute, and, though working with gliders, she was to see the results of her tests applied to military aeroplanes, for soon after Hitler came to power in 1933 Germany had no scruples about building up its Luftwaffe. The Führer soon gave Hanna the title of honorary Flugkapitän, which she was proud to use, in the knowledge not only that she was the sole woman to have received such recognition but that she had done so at a time when Hitler discouraged girls from having careers. The Nazi propaganda machine was working all out to remove women from industry (so that unemployed men could take over their jobs) and encourage them to concentrate on breeding ('to preserve the blood and to propagate the race'). Medals for motherhood (a bronze for four or five children, a silver for six or seven, and a gold for eight or more) were being handed out wholesale.

Young men who had trained on gliders became pilots in

the new Luftwaffe and, like them, Hanna quickly learned to fly powered aircraft. General Ernst Udet, the First World War flying ace, involved her in the direct testing of military aircraft from 1937. She seems to have eased any qualms she might then have had about becoming part of the German war machine by rationalising that she was helping to protect her beloved country, and that German rearmament was not an aggressive action but something which would establish 'a just peace'.

Hanna's most celebrated pre-war achievement was to pilot the first entirely successful helicopter in the world, the Focke-Wulf Fw 61 twin-rotor helicopter. Not surprisingly, perhaps, many other countries, suspicious of Hitler's re-armament and re-formed Luftwaffe, failed to share Ger-many's enthusiasm for this new and revolutionary aircraft. Hanna was instructed by Udet to demonstrate its marvels in 1938 in the Berlin Deutschland-Halle as part of a pro-gramme which had as its theme the loss of Germany's colon-ies (hardly a felicitous one from an international viewpoint). She played her part gamely, although she felt that her dis-play was cheapened by being part of a lavish, circus-type entertainment which was expected to be extremely well attended as it was planned to coincide with the 1938 Berlin Motor Show. With her usual meticulous precision, Hanna mastered every aspect of flying the helicopter and over a three-week period in the Berlin Deutschland-Halle she demonstrated vertical take-offs, sideways manoeuvres, hov-ering and near-vertical landings with extraordinary vigour, ending her performance with the flourish of a Nazi salute.

Although working as a military test pilot in the 1930s and throughout the Second World War, Hanna technically retained her civilian status. Nevertheless Hitler awarded her the Iron Cross Second Class in 1941 after she had performed several successful experiments with a device for cutting the cables of the barrage balloons which impeded the action of German raiders over London. At this time her reputation was at its height. Goering, the Commander-in-Chief of the Luftwaffe, had conferred on her a special gold medal, and her home town of Hirschberg in Silesia gave her the Freedom of the City, as well as a Grünau Baby glider.

As the war progressed, she undertook the testing of ever more hazardous projects, culminating in the piloting of the Messerschmitt 163, an experimental rocket plane which 'took off with a roar and a sheet of flame', reached a speed of over 500 miles per hour in just a few seconds and climbed to a height of 30,000 feet in one-and-a-half minutes. Hanna's first four tests of this machine were successful, but when she took it up for the fifth time disaster struck. The undercarriage, which should have been jettisoned soon after the plane became airborne, jammed. She could have baled out but, hoping to save the aircraft and its instruments, attempted a landing, and crashed.

Shaken but not in pain, she discovered that blood was coming from her head – in fact from an open hole where her nose should have been. Despite this appalling realisation, she made notes of what had happened to the plane before it crashed; she then passed into unconsciousness. Her skull had been fractured in six places, several vertebrae were smashed and her brain had been badly bruised. For some months her life hung in the balance. According to one distinguished brain surgeon, she could not be expected to survive: certainly her flying days must be over.

However, with the determination and resilience that were characteristic of her, Hanna made a partial recovery and, with her nose and face extraordinarily well reconstructed, left hospital. Not surprisingly she had become the victim of almost constant headaches and giddiness, yet, after several weeks of exercises, mountain-walking and a self-established regime of retraining her mind and coordination, she was able to resume her 'normal' extremely challenging flying duties.

In February 1944 Hitler conferred on Hanna the Iron Cross First Class for her work with the Messerschmitt 163. Her mother and several friends were concerned about this, as it underlined her reputation as a Nazi and a militant heroine. By this time, of course, it was evident to even the most patriotically inclined Germans that the war was going badly for the Fatherland. Hanna was distressed to realise, on a visit to Goering, that his grip on reality was somewhat slender. Tortured by the prospect of her country again facing defeat, she then involved herself with a project begun by Heinrich Lange of the Luftwaffe and Dr Theo Benzinger, head of Berlin's Institute for Medical Aeronautics. This was the development of a human flying bomb, intended to damage the Allies' capabilities at key points and to bring about a successful conclusion of the war for Germany.

The Fatherland's 'ultimate weapon' would be manned, directed and exploded by volunteers who would be prepared to sacrifice themselves to save their country from the humiliation of defeat and occupation by the Allies – in Hanna's words, to 'ensure some kind of future for their children'. There were supposed to be plenty of willing volunteers – although subsequent figures dispute this – but those in high command, including Hitler himself, were doubtful about the project. Nevertheless Hanna, who put the idea to the Führer with courageous persistence when he invested her with her Iron Cross First Class, was given permission for its development to go ahead. Although unenthusiastic about the suicide weapon, Hitler seemed prepared to consider it as a last resort, and wanted it to be ready should he ever make the decision to put it into use.

In retrospect it seems extremely unlikely that, at that stage of the war (early 1944), any such weapon could have reversed Germany's fortunes, but work went ahead and there were signs that Nazi Party high-ups were warming to the idea. Goebbels saw its propaganda value in terms of the sacrifices which individual Germans were ready to make for their country, while Himmler, the notorious head of the Gestapo, came up with the bright suggestion that neurotics and criminals should volunteer as suicide pilots, and thus redeem their honour.

Hanna, of course, offered to test the prototype human bomb but, on account of her officially civilian status, she was turned down and military pilots were given the job. Two were severely injured: she and Heinz Kensche, a Luftwaffe staff engineer and pilot, then shared the difficult and very dangerous testing programme. Eventually the bomb was ready to be put into production and operation, but it was far too late. The 'Second Front' had opened early in June 1944; the British and Americans were moving rapidly across France towards Germany, and the Russians were pressing inexorably from the East.

Hanna's involvement with the proposed suicide mission – which she preferred to call 'Operation Self-Sacrifice' – was one factor which helped to tarnish her image abroad, and with many of her compatriots. Another was her presence in Hitler's bunker only a few days before his death. She had been deputed to fly Colonel-General Ritter von Greim (whom she greatly admired, and who may have been her lover) into Berlin on 25 April 1945. Their flight was gruesome. Von Greim was wounded and the aircraft was hit several times when they were attacked by fighter planes and gunfire from the ground. Hanna landed near the Brandenburg Gate and was horrified to witness the widespread destruction of the city.

At the Chancellery, Hitler solemnly promoted von Greim to Field Marshal and Commander-in-Chief of the Luftwaffe. (Goering had by then been arrested as a traitor and stripped of office.) Hanna and von Greim stayed in the bunker until 28 April. She knew that the end of everything important to her was near when Hitler handed her two phials of poison – one for herself and one for von Greim. He told her that he and Eva Braun had decided to take their own lives if Berlin could not be relieved. Two days after she flew von Greim out of Berlin, Hanna heard of the suicides of Hitler, Eva Braun, Josef Goebbels and his wife.

It must have seemed like a veritable Twilight of the Gods for her when, almost at the same time, she heard that all her close family, apart from her brother, had also taken their own lives. (They were the victims of a rumour that all German refugees would be returned to their place of origin. The Reitsch family would therefore have had to return to territory then occupied by the Russians, and they decided that death was a preferable option.) Hanna and von Greim made a suicide pact; he would kill himself first, and she would follow suit (using Hitler's poison phial) eight days after he died. Apparently the delay was arranged so that no discredit might come to him by the inevitable linkage of their names should they die together. In the event, Hanna never kept her share of the bargain. Von Greim committed suicide, but Hanna did not, claiming later that she wanted to remain alive to speak the 'truth' about her country and uphold its 'honour'. She maintained that reports of the gas chambers and other horrors of the concentration camps were lies—deliberate falsifications of the truth by Germany's enemies.

The American authorities, with some persistence, tried to win her cooperation, hoping to draw on her knowledge of German aircraft and rocket developments. She resolutely resisted and stated high-handedly that she would rather be unable to fly, and die in prison, than betray the Fatherland and fly for one of its enemies. The Americans seem to have taken her at her word, for they then held her as a prisoner in Germany for fifteen months.

Several people, including possibly the American Jacqueline Cochran (see chapter 11), suspected, as Hitler's body was never found, that he had in fact been flown out of the bunker in Berlin, and that Hanna might have been his pilot. Hugh Trevor-Roper's book *The Last Days of Hitler* included an eyewitness account of life in the bunker, supposedly by Hanna (but which she repudiated), which suggested a closeness to Hitler and upper echelons of the Nazi hierarchy that added further tarnish to her image.

She was freed in November 1946 and continued to live in Germany although, for the last three decades of her life, she was far from enjoying the status of national heroine which had been bestowed on her in the late 1930s and early 1940s. She felt that rumours about supposed sexual liaisons with Hitler, other prominent Nazis and von Greim were deliberately circulated to denigrate her. Some newspapers published unsympathetic accounts of so-called but fictitious interviews with her, and refused to print rebuttals. She even claimed that anonymous threatening letters resulted in the removal from many bookshops of her 1951 autobiography, *Fliegen mein Leben (Flying My Life)*.

Hanna laid the blame for much of this on the Americans

who, she felt, were vilifying her for her refusal to work with them. Whether she was the victim of a particularly ugly smear campaign, or whether she was simply paranoid, it seems that she was almost driven to taken her own life.

She had the satisfaction of flying again when gliding, forbidden in Germany immediately after the war, began again in 1951. She soon established several new records, but again she felt herself to have been victimised when, as a member of the German team planning to compete in an event in Poland in 1958, she was refused a visa. (Earlier, in 1954, she *had* been granted a visa for Britain to take part with the West German team in the international gliding championship at Camphill; however, the Bonn government made the Aero Club withdraw her from the team because, apparently, some British newspapers were linking her name with Hitler's.) Hanna fell out with the Aero Club when the rest of the team went to Poland without her. She resolutely kept out of German events until she made her peace with the Club only a year before her death in August 1979.

Post-war recognition came to her from countries other than Germany, however, including India (where she took Prime Minister Nehru gliding), Ghana (where she ran a gliding school for several years) and the once-hated America (with President Kennedy receiving her at the White House). Her controversial story has been told in her autobiography and, with insight and balance, by Judy Lomax in *Women of the Air* (1986) and *Hanna Reitsch*, a biography published in 1988.

Even Hanna's detractors cannot deny her courage, resource and truly remarkable aviational skills. Judgment – if it has to be made – rests on whether or not she was more involved with the ideology of Nazism than her duties as test pilot required.

There is no record of Hanna Reitsch inspiring any flying stories for German girls, but it seems likely that W. E. Johns had her very much in mind when he created Thea Hertz in 'The Case of the Black Gauntlet', a story which was included in *Biggles of the Special Air Police* (1953). Thea Hertz is described as having been 'a big noise' in Germany and 'a professional test pilot'. (Hanna was the only well-known female one although Melitta Schiller also tested some planes for the Luftwaffe. However, in the inter-war years, Thea Rasche had become a fairly well-known German stunt pilot. Possibly Johns's Thea Hertz is a fictional combination of Hanna and Thea.)

Thea is very much the baddie of the piece, vindictively determined, several years on, to kill Biggles because he shot down her brother during the war. Her lust for revenge is narratively condemned, and in real life Johns was certainly not generally a person to harbour malice against former enemies. When, for example, Ernst Udet died during the Second World War (ostensibly test-piloting but actually by committing suicide), Johns contributed a graceful obituary tribute to the *Girl's Own Paper*. In his stories, the theme of generosity to former enemies was a recurring one, with female opponents receiving even more chivalrous treatment than Biggles's masculine foes. In his one and only truly romantic involvement, in the story 'Affaire de Coeur', Biggles falls, during the First World War, for Marie Janis, 'a vision of blonde loveliness' who disturbs him so much that he has to hurry back to his quarters, after just talking to her for a little while, on the pretext 'that his magneto is nearly shorting'. She turns out to be a German spy who almost brings about his execution, and does not crop up again in the saga until years later when in *Biggles Looks Back* (1964) she is a prisoner of the communists in Bohemia and has been trans-mogrified from a slightly sexy spy into a long-suffering, dignified and rather passive lady. Biggles and his former German opponent, von Stalhein, cooperate to rescue Marie Janis from communist clutches, and she seems content to let the men do all the planning and take all the risks. The passion that has kept Biggles pretty well virginal for forty-six years and made him (by sublimation) into an expert espionage agent has now settled into an asexual chumminess for Marie which also embraces von Stalhein. Reconciliation between former enemies is so complete that, once they have established Marie in a cosy cottage in Hampshire, Biggles and von Stalhein run down together regularly at weekends to visit her and chat about the stirring old wartime days.

There is nothing chummy, cosy or suggestive of reconciliation about Thea Hertz, but Johns's tolerance of erstwhile
opponents makes his apparent backhander against Hanna
Reitsch the more surprising. Perhaps he, like so many others,
cast her in the role of war criminal rather than patriotic
pilot? In 'The Case of the Black Gauntlet', Thea is certainly a
hard nut. Since the war she has become a 'top-line German
film star'. She and Biggles have been recruited by the Crown
Film Corporation to work on a documentary film which will
be 'the most important air picture since *The Lion Has Wings*
[an actual film made by the Crown Film Unit in 1938 starring
Merle Oberon and Ralph Richardson], for showing at the
International Peace Film Festival at Geneva'.

Biggles is invited to be technical adviser of the air combat
shots, and, rather out of sympathy with all this picture-
making palaver, he snorts: 'Combat? I thought you said this
was to be a peace film?' The producer patiently explains that
the most effective way of 'showing the value of peace' is
probably 'to illustrate the heartbreak of war'. He also points
out that, as the film is intended to be a popular handshake
'between Germany and the Western Powers' Thea Hertz will
fly a Messerschmitt 109 in it in combat with a Spitfire.
Biggles, who is somewhat at sea with the producer's theory
that 'You can't have a film without a girl in it', rather reluc-
tantly accepts the job of technical adviser. He does so partly
in hopes of getting to the bottom of an enigmatic challenge: a
black leather gauntlet decorated with a gold swastika has just
been sent to him anonymously, and he remembers a similar
glove lying near the German plane which he brought down
at Marham aerodrome in Norfolk during the Second World
War. (This, obviously, was the plane flown by Thea's ill-fated
brother.)

Biggles' first sight of the German test-pilot-cum-film-star is
not particularly promising. Her 'slim but rather masculine
figure in flying kit' is aloof yet purposeful, and, on meeting
Biggles, Thea's 'ice-blue eyes' appraise him with 'unnecessary
candour'. When the young male flyer who is supposed to
pilot the Spitfire in mock battle against her Messerschmitt
fails to turn up, she goads Biggles into taking his place.

The producer explains to him that his guns are loaded only with 'dummy tracers that leave a vapour trail which the camera will pick up', but, once in the air, Biggles receives a daunting communication from Thea on their two-way radio connection. Referring to the black gauntlet and the fact of his shooting down her brother, she tells 'Herr Bigglesworth' that she's waited a long time for her revenge: not only is she going to destroy 'this silly childish film', but when she has also demolished Biggles, she will drop her incendiaries on the studios and fly to her friends in East Germany. As a parting thrust she comments, 'The bullets in my guns are real ones,' then abruptly clicks off her radio.

Biggles experiences momentary satisfaction that the mystery of the black gauntlet has been cleared up, and is not particularly alarmed on his own account because he cannot believe that the girl's combat experience will be much of a threat: 'Test-pilotage was one thing: combat-flying was another matter altogether.' He is, however, fearful for the lives which will be lost if Thea drops bombs on the studios, and distressed 'that the woman should have nursed her hatred for an event that had occurred in the normal course of war'.

It is firmly made clear to readers that although Biggles is suffering the severe disadvantage of fighting without weapons against an armed opponent, he didn't feel that it would have made much difference had his own guns been loaded with live cartridges 'for even in those circumstances he would not have used them against a woman'. Bully for Biggles!

There is a breathless dog-fight: 'Round and round they waltzed, with the white chalk-lines made by the tracer bullets cutting geometrical patterns against the blue sky.' But Biggles soon gets fed up with being chased around, diving, zooming and banking. The Messerschmitt and Spitfire come close enough for him to see such a vicious expression of thwarted fury on Thea's face that he is shocked and amazed. Muttering, 'All right my fine lady . . . if that's how you want it, come on,' he slings his Spitfire earthwards like a stone, 'jinks' at ground level to confuse Thea as to his course, and has to duck smartly beneath some telegraph wires. The pursuing German aviatrix is apparently so blinded by hatred that she doesn't see these and, flying straight into them, her plane cartwheels and crashes. Biggles, in his usual 'noblest foe' guise, sideslips down, makes one of the riskiest landings of his career and rushes towards the grounded Messerschmitt which is enshrouded in clouds of vapour caused by petrol running over the hot engine. He knows that one spark will send the whole thing up, and Thea with it. But even as he yells 'Don't shoot!' and flings himself flat, two things are photographed on his brain: 'narrow carmine-painted lips and the black circle of a pistol-barrel', and, as Thea fires in a last desperate attempt to eliminate Biggles, the vicious 'whoof' of exploding petrol engulfs her.

So much, then, for vindictive, German, female test pilots who should have known better.

During the Second World War, the USSR employed several women as test pilots, although this was not widely publicised elsewhere. Critical conditions made it impossible for the Russians to share their allies' scruples about female flyers in combatant roles, and, by the end of the war, girl aircrews had flown 24,000 sorties against the Germans. When Operation Barbarossa (Hitler's unexpected attack on the Soviet Union) began on 22 June 1941, the artillery barrage opened up from the Baltic to the Black Sea, and German bombers, with the advantage of surprise, strafed all the Russian frontier airfields, destroying many squadrons of parked fighters and bombers. Even when the Russian pilots, who had had precious little flying time, *did* manage to get into the skies to attack them, the already well-battle-blooded German flyers had no difficulty in maintaining air superiority. Within a few months the ferocity of the blitzkrieg in the air and on the ground brought Hitler's troops only 20 miles short of Moscow, with Leningrad under siege. The Red Air Force threw everything possible in terms of men and machine-power against the Luftwaffe, in a desperate and successful endeavour to gain time for aircraft factories in the path of the rapidly advancing Wehrmacht to be dismantled and moved to the comparative safety of territory east of the Ural mountains, out of the German bombers' range.

By October 1941 the losses of Russian aircrew members had been enormous. Nevertheless many people were surprised when their country's most celebrated aviatrix, Marina Raskova, broadcast an appeal on Radio Moscow for women volunteers to become front-line combat pilots. The truly overwhelming response brought sackloads of applications, and Marina, with her small band of helpers, sifted from these the names of some 2000 girls who seemed particularly appropriate, and who would be called for interviews. Three women's air regiments (one of fighters and two of bombers) were to be formed; each would have three squadrons of ten aircraft and be 'manned' entirely by female pilots and aircrew, mechanics and armament fitters.

At this time, of course, the concept of all-girl combatant aircrews was indeed revolutionary, although apparently Sabiha Gokcen, the adopted daughter of Kemal Ataturk, President of Turkey from 1923 to 1938, had led a female-crewed squadron of bombers into action in Turkey in 1936 (see chapter 11).

Marina Raskova, a Hero of the Soviet Union, was the ideal person to make the appeal for female volunteer pilots. In September 1938 she had become a national idol when, with two other women flyers, Valentina Grizudobova and Polina Osipenko, she had established a 6000-kilometre world record for a non-stop direct flight by women. The three 'Winged Sisters', as they had been dubbed, had flown from Moscow to Komsomolsk-on-Amur, with citizens from all the Soviet republics listening intently to hourly radio reports of their progress. It seemed that the hopes of the three aviatrices were to be cruelly dashed when, just under an hour's flying time from Komsomolsk-on-Amur, they ran into a terrible snowstorm, when heavy ice coated the wings, and Valentina, who was piloting, could no longer keep the twin-engined Rodina plane at the necessary height.

Everything that could be jettisoned was thrown out, in order to lighten the aircraft, but still it dropped through blinding sheets of falling snow towards the forests below. Suddenly, without saying a word to her comrades, Marina parachuted herself out of the plane. She had been the navigator, and the cross which she marked on the map showing their approximate position at the time of her drop was all that eventual search parties had to go on. It was not until ten days later that a hunter found her, food- and water-less and absolutely exhausted, in bleak, almost impenetrable forest land. Her courageous self-sacrifice endeared her immediately and lastingly to the public.

The young women from whom Marina had to select her first air-crews were not all untried flyers. Her own exploits had already inspired many girl students and factory workers to join local flying groups or the Osoaviakhim, a network of paramilitary flying clubs which had been in existence for some time, providing free instruction in glider and powered flight for girls of seventeen and upwards. Several were already experienced flight instructors, and/or pilots of air-liners operating within the Soviet Union (these must surely have been the first civilian female airline pilots in the world, though this is not often acknowledged).

Marina (by then Major) Raskova warned all the short-listed volunteers that they would have to fight like men. No quarter would be given – the girls might be killed, burned beyond recognition, blinded or severely maimed. They might be captured by the Germans and subjected to rape and torture. But none seems to have been deterred. The achievements of the three all-female air regiments were not fully told (at least to the public outside of Russia) until 1981, after Bruce Myles, a British journalist who was researching the air war on the Eastern Front, stumbled by accident on some obscurely tucked-away paragraphs about women combatants in an official Soviet Second World War history. His gripping book, *Night Witches*, carefully documents their outstanding achievements in the face of tremendous odds.

The girls were in the main young, lively, optimistic and awesomely courageous. At first they were put into hideous, ill-fitting, discarded male uniforms, which they adapted and smartened up with panache and ingenuity. Despite the toughness of their mission, they retained their femininity, wearing make-up, flimsy nightgowns and underwear, and flowers in their caps and tunics. They kept their hair long whenever they were allowed to do so and, while awaiting the order to 'scramble' filled the uneasy hours of inaction by knitting or doing embroidery.

They lived in their all-girl groups with a strong sense of sisterhood and solidarity. Bitchiness and lack of discipline were almost non-existent. They encountered, and overcame, prejudice from senior male officers, and married or had love affairs with the men they flew and fought beside.

While the war was still in progress, the British public received and responded admiringly to impressive accounts of determination and dedication on the part of the Russians. In particular they were moved by reports of the scorched-earth policy, with peasants burning their farms, and workers destroying their factories and homes so that nothing of use could fall into the hands of the advancing Germans. There were also the Stakhanovite efforts of Russian factory operatives, who literally worked all hours to rebuild their airforce after the early massive destruction wrought upon it, and to continue to equip their land and sea forces. Rumours circulated in Britain about female involvement in combat, but these centred on images of butch girls built like tractors slogging it out shoulder to shoulder with their men in the snows or slush of Leningrad, Stalingrad or Moscow. Surprisingly, in view of the euphoria which the Battle of Britain pilots had inspired only a year or so before the USSR's female air regiments became operational, accounts of girl flyers rarely filtered through, either as rumour or in reports by radio and the press.

However, Dorothy Carter, the author of the already well-established series of Marise Duncan's flying adventures, seems to have been aware of the Russian airwomen's contribution to the war effort. Her *Comrades of the Air* (1942) sends her ever-active aviatrix heroine into Russia. She wants to transport greatly needed aircraft to the Red Air Force and is somewhat vaguely instructed to find out what else the population needs – 'clothes – food – mobile canteens', etc. (One would have thought that the Red Cross and Mrs Churchill's Aid to Russia Fund could have drawn on more comprehensive information about this than whatever a youthful aviatrix could glean.) Nevertheless, Dorothy Carter's stories convey a sense of authenticity, and this is spiced by the occasional introduction of real-life characters and situations. Marise is determined to get to Russia because her father and her former colleagues ('battle-scarred Jim Grant, one-eyed hero of an earlier war' and former flying instructor Jim Custance and Tony Arcoll) are off there on some secret mission.

She approaches her chief, 'the woman who was responsible for the network of women pilots who, all over England, were engaged on the same work as Marise'. Though not named, this character is obviously Pauline Gower, the leader of the women's section of the ATA. More than sympathetic with Marise's request to ferry aircraft to Russia, she agrees that 'although we are already shipping' planes in considerable numbers, 'to fly them there would be quicker'. Marise, apparently, is to pioneer this air-ferrying service to Russia.

In fact this was probably impractical, unless there were refuelling facilities *en route*, but Dorothy Carter glosses over how Marise actually flies into Russia and finds the airfield on which she has to land. She is simply and suddenly established in a women's air force dormitory bed somewhere in Russia after 'a long and gruelling flight'. Incidentally, before she embarks upon her assignment, her woolly but well-meaning mother warns: 'Take care of yourself in Russia. They've some funny customs there, I dare say. I believe they all call each other Comrades, or something,' and 'the Russians are strange people. I've got a book somewhere – Tolstoi – not very cheerful.'

Ignorance about our fairly new-found allies was not, at that time, confined to Mrs Duncan, and Dorothy Carter sets out deliberately to dispel certain misconceptions and to 'humanise' the Russians – particularly their women – for her readers. Marise is immediately on chummy ('comradey') terms with her dorm-mate Katya Petroff, a tremendous patriot and an air-gunner who takes pride in her performance: 'I have shot down two Heinkels already this week.' Replying that this 'sounds grand', Marise has doubts about whether she would like to attack a German plane, but these are soon resolved. With the Wehrmacht advancing, Katya's group of flyers have to abandon and completely destroy their airfield, and take off to another. Marise, piloting a 'sweet to fly' Russian bomber (probably a Tupolev SB-2), transports to safety air-gunner Katya, her navigator brother Ivan and Marise's injured-in-an-air-raid father, Captain Duncan, shooting down a German fighter on the way. Katya is impressed – and even more so when, after force-landing in snowy regions behind the German lines, Marise's knowledge of aircraft maintenance turns out to be greater than Katya's or Ivan's. Off the coast of the White Sea, Marise is alerted by ice-floes drifting on the water to drain the fuel off from the engine before it freezes solid. She also shows her companions how to build a shelter out of fir branches. (She is, of course, in both instances drawing on knowledge gained from her Arctic Canadian adventures in *Snow-Queen of the Air*.)

Katya and Ivan compliment Marise on her cleverness, and she chuckles that not long ago they were pitying her because she hadn't done any aerial fighting. Katya admits that, on hearing British WAAF and ATA girls were 'taking work to release men for the fighting', she had thought they must be feeble creatures compared with the Russian girl combatants. However, now she sees that she was wrong. In fact it is Marise, full of British phlegm and breeziness, who becomes the leader of the party as they struggle to get back into Russian-held territory. She and Katya have helpful chats on the way about the stereotyped images which the British and Russians have of each other's country.

There are many echoes in the story of experiences of the real-life members of the women's air regiments. *Night Witches* gives a graphic account of two Russian airwomen, navigator Ira Kasherina and Sonya Azerkova, who, like Marise and Katya, have to evacuate an airfield and then find themselves trapped behind enemy lines. Dressed as peasant women, they encounter two German soldiers on a stretch of lonely road, who try to rape them – and get much more than they bargain for. Ira shoots one in the face and Sonya shoots the other in the head. Both women wanted to be sick: killing the enemy on their night-flying missions was acceptable, but shooting them at close range on the ground was very different.

Grounded Russian female flyers came across civilian girls who had been subjected to sexual atrocities by the occupying German armies. *Comrades of the Air* reflects these only partially for it is, after all, primarily a book for ten- to fourteen-year-olds. Marise, Katya and Ivan find themselves in the position of having to protect innocent women and children from the gratuitous violence of the enemy, and, if they have to kill the occasional soldier at close range, they suffer none of the nausea which gripped Ira and Sonya in similar but real-life circumstances. It is significant that when Marise brings down her first German fighter, she is dazed but cheerful, if not actually elated. In reality baptism of fire for the women of the air regiments often involved retching, shaking and 'sheer terror'. There was certainly none of the exuberance that Katya experienced when shooting down her two Heinkels and boasting to Marise about their destruction.

Marise and Co. do a lot of wandering around in icy wastes, where they stumble upon a secret U-boat base, before the Germans catch up with them and start to transport them by lorry to Berlin (a long way off) for interrogation by the Gestapo. Their (fictional) inefficiency and slowness seem to support Marise's earlier suggestion that 'the Nazis were a lazy lot of blighters, really. They ought to have gone to look for us at once,' but, of course, real-life Russian air girls on the run experienced no such shilly-shallying on the part of the Germans. Female flyers would stay in damaged aircraft until the last possible moment, even at the risk of suffering terrible burns, in the desperate hope of being able to bale out over their own lines rather than fall into enemy hands.

Nina Karasova, a teenage navigator, was captured, interrogated and ultimately incarcerated first at Ravensbruck and then at Buchenwald concentration camp. Eighteen months later, when the guards deserted Buchenwald as the Allied armies approached it, the unrecognisable Nina, barely alive after terrible degradation, looked an old and agonisingly emaciated woman.

Marise, Katya, Ivan and Captain Duncan are more fortunate. Marise manages to escape from the lorry bound for Berlin. Astoundingly but felicitously, she runs into Jim Grant who has just arrived to airlift them all with remarkable speed and adroitness to safety: 'Jim, isn't it amazing? Here we are, alone in Russia, with thousands of Jerries just across a field or two, and we're chattering amiably of this and that!' Marise flies her injured father home to England, but first she swears to Katya and Ivan that she will be back, bringing them more planes. ' "And we will fight again, shoulder to shoulder," said Katya fiercely. "Allies," said Ivan, holding out his hand. "Allies," smiled Marise, shaking it.'

Later she confides to her mother over cosy afternoon tea by the log fire burning in the grate of their English village home: 'Well, I've fought beside them, and I've been through some pretty tough times with them, and . . . they are Comrades of the Air.'

It is interesting that the still-shadowy figure of Dorothy Carter seems to have been the only writer to attempt to bring home to British readers at least some measure of the achievement of the Russian women combat pilots. It was almost forty years on before we had access to their true stories for the first time. Amongst the most indelible images of bravery must be that embodied by Lily Litvak, 'The White Rose of Stalingrad', and truly 'the pride of her regiment'. This pretty, blonde and popular young fighter-pilot passionately loved and despairingly lost her fellow flyer, Alexei Salomaten, when he crashed. She marked her Yak fighter with a rose for every enemy 'kill' that she made, and struck terror into the Luftwaffe – 'Achtung! Litvak!'

In the end, after twelve months of intensive combat duty, Lily was killed when two Messerschmitt 109s brought down her flower-adorned fighter. The exploits of 'The White Rose of Stalingrad', and those of her comrades are reminders of one of the most popular (but rarely since reprised) songs of the Second World War. Its refrain melodiously promised, in those terrible days when German troops occupied large tracts of the country, that 'You will surely bloom again, my lovely Russian Rose'.

Chapter Eleven

The Fastest Women in the World

KITTY'S DONE 1,700 HOURS FLYING — 500 AS A CAPTAIN.

JEAN'S BEEN SENDING RADIO MESSAGES IN A ROYAL STUARRRT ACCENT FOR 1,500 FLYING HOURS.

PAT'S NAVIGATED EVERY ROUTE FROM HERE TO ISTANBUL WITHOUT LOSING HER MONOCLE, LET ALONE HER WAY.

AND I'VE PUT 500 HOURS IN MY BOOK AS KITTY'S 2ND OFFICER— APART FROM 1,200 BEFORE.

Kitty Hawke and Her All Girl Air Crew, in *Girl*, 1951.

I might have been born in a hovel but I determined to travel with the wind and the stars.

JACQUELINE COCHRAN
The Stars at Noon (1954)

THE Second World War officially ended in 1945, but the peacetime era did not really begin until the early 1950s. During the late 1940s the world was still licking its deeply scored wounds; hunger, refugee and population displacements, power struggles between supporters of opposing ideologies and bitter arguments about relocated national boundaries were rife. On the one hand the USA's Marshall Plan promised help to slowly recovering European countries; on the other, the beginning of the Cold War suggested that the longingly anticipated 'lovely day tomorrow' of wartime popular song would never come. The international outlook was bleak, with little suggestion that the political and social freedoms for which so many people had fought and died would be realised on anything like a universal scale. Above all, of course, despite the deterrent factor of the nuclear bomb, the unyieldingly rigid East–West divide provided a constant reminder that a future – and even more horrendous – world war might not be far away.

By 1950 many of these challenges were still unresolved, but human resilience and optimism were once more asserting themselves. In Britain the new boys' weekly paper, *Eagle*, launched on 14 April, caught the expansive mood, encouraging readers to feel that the world was their oyster, and that man's extraordinarily advanced technology would be intelligently and positively harnessed. The paper was the brainchild of the Reverend Marcus Morris, a Birkdale vicar who, disturbed by the importation of American horror comics into the British market, made up his mind to produce a quality boys' magazine which would embody Christian ideals. Chad Varah, then a vicar in nearby Blackburn (and who became well known as the founder of the Samaritans), cooperated in the project, which really sprang to life when the recently demobilised artist, Frank Hampson, became involved in its planning stages. The *Eagle*'s blend of muscular but imaginative Christianity worked well. Its overtly religious picture-strip features such as 'The Great Adventurer' (the life of St Paul) were robustly presented; so, too, were the paper's strip and text stories which featured a series of charismatic personalities, from the upper-crust policeman Archibald Berkeley-Willoughby (PC 49, an already established radio favourite) to the comic pirate, Captain Pugwash, and bungling sleuth, Harris Tweed.

Basically, however, it was Frank Hampson's brilliantly drawn cover picture story, 'Dan Dare, Pilot of the Future', which made *Eagle* truly addictive. Neither the Russians nor the Americans had got into space by 1950, although the race was on and everyone felt that it was only a question of time before one or the other did so. Dan Dare blazed the thrilling fictional trail in advance of real-life achievement. (His exploits were actually set in the first part of the twenty-first century, but he was quintessentially of the early 1950s.) Grown men and women, as well as the million or so young boy readers for whom the paper was originally designed, bought it avidly each week, and echoed the slogan that so often adorned the front page: 'Will Dan Dare reach Venus?'

Eagle represented a revolution in the juvenile comic and boys' paper. At a time when magazines were bereft of new ideas and still reflective of the austere wartime mood, its twenty-page tabloid colour-gravure format was a triumph of design and intriguing contents. The sense of action conveyed by its breaking-out-of-their frames pictures echoed the enhanced visual awareness of the first generation of regular juvenile TV watchers. The illustrated strips, and also the text stories, managed to suggest satisfying intricacies of character and relationship, while the inventive use of space technology in the Dan Dare adventures was so much appreciated by readers that *Eagle* also offered vividly detailed 'exploded' cut-away pictures of current rockets, aircraft, ships, engines and gadgetry.

Nevertheless, despite its air of modernity, *Eagle* still bore traces of the exclusively male dream-world of the pre-war *Boy's Own Paper* and, with the notable exception of Professor Jocelyn Mabel Peabody (see chapter 13), had no use for girl characters in its stories. In spite of this, girls as well as boys read *Eagle*, so, not surprisingly, in the year following its launch, the publishers, Hulton, decided to bring out a femininely slanted paper, appropriately named *Girl*. It began on 2 November 1951 and continued until 3 October 1964. (The title was relaunched in February 1981, but with such a different format that the new paper had no recognisable links with the old.)

The stars of the show in 1951 were to be Kitty Hawke and her all-girl air crew who, like Dan Dare in *Eagle*, were featured in lively picture-strips occupying the whole of the front cover page and its reverse. The artist was Ray Bailey, whose illustrations cannot be faulted. However, Kitty Hawke and Co. never got too far beyond the launching pad, with their adventures running for less than a year.

Blonde and appealing Kitty is a natural leader with a well-balanced personality. Her crew includes the Eton-cropped navigator, the Hon. Patricia D'Arcy, who sports a monocle and seems to hark back to the Wodehousean heroines of the 1930s' girls' papers, Jean Stuart, the Scots radio operator, and Second Officer Winifred ('Windfall') White, a tiny toughie who wears her uniform cap at a pushed-back angle that makes her look butch and vaguely reminiscent of the GIs who had been familiar sights to many English girls only a year or so earlier. Kitty's father owns the Hawke Air Charter Co., which employs them – but there is no question of any nepotism. Early in the story Kitty and her crew, like so many fictional and real-life flying heroines who preceded them, bemoan the fact that they only get the duff assignments although they are extremely experienced ('Kitty's done 1700 hours flying – 500 as Captain,' Windfall's 'put in

500 hours as Kitty's Second Officer – apart from 1200 be-fore', Jean's been 'sending radio messages . . . for 1500 flying hours' and 'Pat's navigated every route from here to Istanbul without losing her monocle, let alone her way'). Plum jobs, such as ferrying the stars in a film unit to North Africa, go to Smedley, the male senior pilot, or Carstairs, a stereotyped ex-RAF type: 'After all, gels are only gels, aren't they? In an emergency one needs the steadier qualities of the male, what?'

Of course, despite all this predictable male prejudice, Kitty Hawke and her chums soon find themselves at the centre of the action in air-sleuthing adventures which involve kidnap, hidden treasure and the restoration of a rightful king to his throne. ('One more job chalked up to the all-girl aircrew to prove to Dad that we can operate his planes as efficiently as the glorious males!') Stirring but predictable stuff – which regrettably included none of the addictive inter-terrestrial elements that spiced up the Dan Dare strip. The most excit-ing thing about the short-lived Kitty Hawke story was simply that it featured an all-girl air crew. This seemed to strike a wildly progressive note in 1951 at a time when few young girls expected to drive cars, let alone to pilot or navigate aeroplanes.

Readers of *Girl* in 1951 probably knew nothing of the involvement of Russian all-girl crews (and indeed all-girl regiments) in aerial combat during the Second World War (see chapter 10). Nor were they likely to have heard of an even earlier all-female squadron which had flown – and seen action – as long ago as in the late 1930s. Fact had thus dramatically presaged fiction in the career of Sabiha Gokcen, the first Turkish woman pilot. Born in Bursa in 1919, she was orphaned in childhood and adopted by the President of Turkey, Mustafa Kemal Ataturk, in 1925. Sabiha was edu-cated at the American Girls' College in Istanbul, and after-wards graduated from the Turkish Aviation School in 1935. She received further training in gliding in the USSR and at Eskisehir Military Air School, and, from 1936, took part in military exercises and led a female squadron. It seems curious that the first fully functioning female military pilot should have come from a Moslem society – indeed, as a story-line it might have been rejected as unbelievable.

From 1938 Sabiha travelled in what seems to have been an unofficial ambassadorial role to the capital cities of various Balkan states, piloting her own aircraft. Later on she became an executive and a teacher at the Turkish Air School until her retirement in 1955. Little else is known of her outside Turkey today, although in December 1941 the *Girl's Own Paper* published an engaging photograph of her in 'Women with a Will' by Sam Bate. This article mentioned that although 'the emancipation of women was not popular' in Turkey, Sabiha had fought hard for better conditions for her sex, had driven a high-powered car 'like a man' and become 'not only the First Turkish Airwoman, but Turkish Air Ace Number One'. Apparently, 'From frowning upon her Western ways, the Turks slowly began to admire her, until she became the most popular woman in the country.'

Possibly because Kitty Hawke failed to appeal widely to *Girl* readers in 1951–2, all-girl air crews did not crop up again in fiction until the 1970s, when they became features of the Space Age. When Kitty and Co. flew off into the sunset in 1952, their cover spot was taken over by Wendy and Jinx, Schoolgirl Detectives (drawn, as Kitty had been, by Ray Bailey). Though reasonably appealing, this pair of heroines offered nothing to *Girl* readers which had not been served and rehashed many times over during the 1930s and 1940s. Nevertheless they seem to have provided what was wanted, and became long-standing favourites in the paper.

The 1950s, in fact as distinct from fiction, were marked by outstanding achievements by women flyers. In particular, America's Jacqueline Cochran became the first woman in the world to fly faster than the speed of sound on 18 May 1953. She not only crashed the sound barrier but by sheer determination and force of personality pushed out other frontiers of female aviation. She was both an outrageous individualist and a person who cared passionately for her country and for fellow-flyers, even though her expression of these concerns was often abrasive, aggressive or embarrassing.

Although the 1950s saw the heightening of career ambitions for women pilots, there was still an acceptance by many of them that they would always only play supportive roles. Ruth Butler, for example, had learned to be a flight mechanic while still at High School in Maine in the 1940s. She made her first solo flight in 1952 and a year later became the first woman to fly, as part of an expedition, over both the North and South Poles. She served the group as a flight technician. However, her aeronautical ambitions fell short of the top jobs: 'I don't want the glamour of stunt-flying or racing or flying jets. I think it would be wonderful to be an airline pilot, but women haven't broken into the field yet. I think the ideal job for a woman flyer would be as secretary–pilot to a business executive, flying his private plane and accompanying him on his trips' (*Operation Polar*).

Jacqueline Cochran refreshingly refused to set her sights lower than those of any man. She frequently commented that she wasn't interested in setting new women's records (though she often did) but worked to set absolute ones. She regarded even prestigious female events, such as America's All-Woman Transcontinental Air Race, as second-best, and felt that by competing with men she became part of the flying elite, and opened doors for the relationships which she so much valued with flyers Chuck Yeager and Fred Ascani, several air force generals and President Eisenhower. She saw herself as an equal member of this kind of unofficial club, and – somewhat irritatingly for other female flyers – tended to associate with women's groups and events solely in the role of benefactor or adviser.

Jacqueline had become president of the famous association of women pilots, the Ninety-Nines, in 1941. By 1953 she had secured all but one of the principal world speed records for straight-away and closed-course flight. Her awards included the Harmon trophy (1938 and 1939), which was presented annually to the most outstanding American female aviator; the Distinguished Service Medal in 1945, which was generally conferred only on members of the armed services; the French Légion d'Honneur, and various Woman of the Year awards. In 1959 she was elected president of the Fédération Aéronautique Internationale; she also received several honorary literary degrees, which must have given her especial satisfaction as, according to Judy Lomax in *Women of the Air*, when she obtained her pilot's licence she had to take 'the usual written examination orally as she was still barely literate'.

Jacqueline Cochran's life embodies an archetypal rags-to-riches story, although the Cinderella figure at its centre is far from being a passive, promoted-through-her-beauty-and-sweetness character. An orphan of unknown parentage, Jacqueline was born somewhere between 1906 and 1910 (her accounts of her birthdate are vague) and brought up as the adopted daughter of a feckless and extremely hard-up couple in a sawmill town in Northern Florida. She tried unsuccessfully in childhood to escape from this Tobacco (Sawdust) Road atmosphere, first with a group of circus performers, then by trying to join some gipsies. Her 1954 autobiography, *The Stars at Noon*, graphically records her early life, when, garbed in old sacking, she had to sleep on the floor, and, to supplement the inadequate diet provided by her adoptive parents, scavenge for crabs and clams. Around the age of eight or ten, in gross violation of the labour laws, she went to work in the sawmills, doing a twelve-hour night shift for 6 cents an hour.

The bleakness of this period of her life was alleviated to an extent by the 'marvellous influence' of Miss Bostwick, a teacher who during Jacqueline's rare periods of schooling had encouraged her to learn to read (this seems to have been an uphill battle for Jacqueline, who nevertheless worked at it for 'several hours a day' whenever she had the chance). She soon realised that while working in the sawmills she would be unable to pull herself high enough by her own boot-straps to build a better life. She left home, ambitious to earn sufficient to buy some decent clothes and a car – and to travel. The ferocity of determination which was so characteristic of her carried her through a series of jobs in beauty shops and hairdressing, and enabled her to buy her first car – a Model T Ford – before she was sixteen. This was her very important 'first experience with engines . . . I personally ground the valves.' She took up and completed a three-year nurse's training, but after a short spell of work with a doctor in a sawmill town she abandoned any ideas of a career in the medical world: ever a realist, and depressed by the sickness and misery of the patients, she felt that she could never help anyone substantially unless she had money.

Returning to the hair and beauty business, she was taken on in the early 1930s by the famous coiffeur Antoine, and worked at both the New York and Miami Beach locations of his Saks' Fifth Avenue salon, following the seasonal migrations of her affluent clients. Many of these invited her to smart parties, at one of which, in 1932, she met Floyd Odlum who, several years her senior, was a millionaire and Wall Street financier running Atlas Utilities and Investors Co. Ltd. Self-made like Jacqueline, but – unlike her – quiet and serious, Odlum was very much attracted to her. At their first meeting, he suggested that she needed wings to implement her elaborate and ambitious programme of going on the road to set up her own cosmetics company. She immediately pursued this idea, spending her holiday learning to fly at Roosevelt Field, Long Island, and gaining her private pilot's licence in just under three weeks, undeterred by accidents and disasters which she witnessed during her training: 'As I was about to take off, a plane spun in across the field and my flight was delayed while the fragments of man and plane were being collected.' On the following day, she saw 'some bodies' being dragged out of the marsh at the end of a landing-field.

Jacqueline obtained her commercial licence in 1933 and subsequently continued to combine cosmetics and aviation careers. She set up a beauty salon in Chicago and a laboratory in New Jersey and flew regularly between them. She began to enter air races and was soon breaking record after record. In 1938 she became the first woman to win the prestigious Bendix Trophy race in a Seversky pursuit plane which she had 'never flown' until the night of take-off. She had to overcome a serious problem: the fuel pipe was partially blocked by some paper, making it necessary for her to fly the race with one wing higher than the other in order to maintain an adequate petrol supply. Always liking to display a 'bandbox-fresh appearance' after any event, she made the judge of the race, who came to meet her on the runway, wait for some time while she repaired her make-up.

Jacqueline and Floyd Odlum were married in 1936 and they lived in great and glitzy style at the Coachella ranch in Southern California, entertaining influential luminaries from film stars to duchesses, high-ranking army officers and presidents. Floyd was a friend of the Roosevelts and other prominent politicians. Soon after the outbreak of the Second World War, and two years before America came into it, Jacqueline wrote to Eleanor Roosevelt about training and using women pilots to release male flyers for combat duties. She received a sympathetic ear, although no such programme was immediately launched. She was determined to get into the war somehow and managed to persuade General Harry ('Hap') Arnold and Britain's Minister of Aircraft Production, Lord Beaverbrook, to let her fly a bomber across the Atlantic. Although Britain needed every plane she could get, this seems to have been mainly a propaganda and self-promoting exercise on Jacqueline's behalf. Military pilots resented a civilian – and female – flyer having control of a bomber, so she had to agree to a military pilot travelling with her to make the take-off and landing.

In London she met Pauline Gower, who suggested that Jacqueline might be able to recruit American women pilots to augment the women's section of the ATA. She hoped for some 200, but, in the event, after meticulous research, interviewing and correspondence, Jacqueline produced a list of twenty-five thoroughly experienced, suitable and willing recruits. One failed to pass the medical examination, but the other twenty-four arrived in Britain fairly early in 1942 to take up their ferrying duties. Flying in to welcome the first batch, Jacqueline was given the honorary title of Flight Captain, which was resented by some of the ATA members on the grounds, that, unlike them, she was doing very little flying at this time. Nor did she endear herself to them, during this period of austerity and strict rationing of food, clothes and fuel, by frequently turning up at airfields in a

Rolls-Royce and sporting a mink coat (the fact that this was only her second-best probably made matters worse).

It was some months after the December 1941 attack by the Japanese on Pearl Harbor, and the USA's entry into the war, that a programme of training for American women pilots was implemented. Jacqueline returned in the autumn of 1942 to find that Nancy Harkness Love (a well-known pilot who had, as long ago as 1940, helped to ferry planes for Britain and France to and across the Canadian border) was already organising a Women's Auxiliary Ferrying Squadron as part of the US Army Air Force's Transport Command.

Jacqueline felt that promises about command positions made to her earlier by General Arnold and others had been betrayed, and, with her usual persistence in haranguing those at the top, she brought about a situation which gave her as much control as Nancy Love, if not more. Jacqueline was to be responsible for training, while Nancy continued to organise the ferrying work. Soon their two groups were merged to become the Women's Airforce Service Pilots (WASPs), with Jacqueline as Director of Women Pilots and Nancy as their executive member of the staff of the ferrying division of the Air Transport Command.

Jacqueline's wartime experiences plunged her ever more deeply into the worlds of the rich, the famous and the influential. At the ending of hostilities she visited the Pope, presidents and royalty – as well as many of the most notorious areas of operations, including Buchenwald, the Nuremberg courtroom during the German war criminals' trials (where she started a minor riot when one of 'Hitler's cohorts' showed pride in having been a prominent Nazi) and the Chancellery in Berlin. Loftily sweeping aside the account of Hitler's suicide, because she was convinced he hadn't the guts to take his own life, she noted the landing-strip in the grounds, and inclined to the view that he had been flown out to safety (one wonders if she suspected Hanna Reitsch's involvement in such a rescue flight). She was unable to see Hiroshima but visited devastated Tokyo, as well as Shanghai, India and elsewhere. Her travels, as described in her biography, endorsed in a series of uneasy clichés her appreciation of the American way of life, her doubts about tolerance towards former enemies, and her conviction that God was firmly supportive of her own entrepreneurial attitudes.

She hoped for nomination as a Republican candidate to Congress, but ironically was instead twice approached by the Democrats – whom she turned down. Flying again, she met and became friendly with Captain Charles ('Chuck') Yeager, the American military pilot who was the first person to break the sound barrier in 1947. Chuck coached her for her own attempt to fly faster than the speed of sound, and, in fact, probably saved her life on one of her low-level flights when he noticed a fuel leak and pointed it out to her. On 18 May 1953 Jacqueline broke the sound barrier in a Canadian F86 Sabre. On the same day she established a world speed record for women of 652 miles an hour. (She broke three men's records, too).

Despite her psychologically robust and even aggressive approach to life, Jacqueline's health was not of the best. Stomach, sinus and optical problems involved her in at least twelve major operations. In her sixties she was forced to abandon competitive flying because of a serious heart condition. Floyd died in 1977, and Jacqueline declined fairly rapidly. Coachella ranch was sold and, irritated by her physical weakness, she moved into a fairly modest house. During her last years she suffered a great deal of discomfort, with her body so heavily swollen by kidney and heart disease that she had to sleep sitting upright in a chair. Few of her former colleagues and friends visited her – she had, of course, made some enemies in the course of her long career in aviation. During the Second World War her bossiness had upset various male military high-ups; her disparaging report about the WAF (Women in the Air Force) in 1950 (of which she knew little at first hand) alienated its officers, and her lack of support for female flyers who hoped for selection as trainee astronauts for the Mercury Space Programme in the 1960s brought about further resentment.

It seemed a sad end for someone who had been the fastest woman in the world. Her funeral in 1980 was attended by only fourteen people. Curiously Jacqueline Cochran, a quintessential high-flying, go-getting, rags-to-riches American, proved too uncomfortable a figure to be fully appreciated in her own country.

For the twelve years preceding 1963 her records were under constant challenge from a European aviatrix, France's Jacqueline Auriol. Starting to fly several years after the American, the second Jacqueline, like the first, had to dig deep into her reserves of courage and resilience in her aeronautical career.

She met Paul Auriol before the Second World War, when they were both studying in Paris. Jacqueline's subject was art, and Paul's political science. Neither of their families approved the match: Jacqueline's were firmly traditionalist and Catholic, while Paul's father, Vincent, was an atheist and a militant socialist. Nevertheless, with predictions of war seeming likely to become a reality, the two families permitted Jacqueline and Paul to marry early in 1938.

After the fall of France to the German armies, Paul's father made no secret of his opposition to the Vichy government and, after a period of internment, was allowed home under close supervision. In spite of this the Auriol family linked up with the Resistance, and Jacqueline frequently found herself fleeing from the Germans with her husband and sons. Their fortunes changed after the war, when Vincent Auriol, after several political triumphs, became France's President in 1947. Success brought its problems, however. A campaign to smear the President's image rubbed off on to Paul and Jacqueline, whose wealth was questioned and who were said to be involved with the black-marketeering of jewellery, penicillin and drugs.

Recreational flying was originally an escape for the couple from the innuendo to which they were subjected, and, although Jacqueline was at first only mildly enthusiastic at being airborne, she became addicted as her piloting skills improved. She studied aerobatics with one of France's leading exponents, Raymond Guillaume, and, flying in her own right, began to give aerobatic displays. Tragedy struck when, on a publicity flight with Guillaume and with someone else piloting, their SCAN 30 amphibian aircraft crashed and Jacqueline suffered appalling head injuries, although Guillaume and the pilot were relatively unhurt.

Jacqueline's life was in the balance, and her face had suffered permanent disfigurement. Cosmetic surgeons were summoned but felt that little could be done to mend the face of the woman whose beauty had once been a byword. The many fractured bones refused to knit; the pulpy flesh did not heal, and, in despair, Jacqueline wouldn't allow her small sons to see her for the best part of two years. The lethargy that continued to engulf her was briefly relieved when Guillaume took her on a flight, during which she handled the controls and knew that at least she could still fly. To test her doctors' theory that her face would not improve because she had no wish to recover, Jacqueline spent some time in a mountain retreat, with various aeronautical textbooks and a human skull: she hoped that the latter might enable her to learn more about her inability to heal. She also took a loaded gun, meaning to kill herself if it seemed that she truly lacked the will to recover. Study and solitude convinced her that she had no desire for death or permanent invalidism. She embarked upon three months of difficult but reasonably successful facial operations and, though still disfigured and shy of public attention, began to fly again. Then a series of operations in America gave her a new nose and cheekbone, and an acceptable face once more.

Her plan was to become a solo test pilot but, to achieve this, she needed to prove her special capacities and dedication. She did so in 1951 by breaking Jacqueline Cochran's 441-mph record, which had stood for four years. The American generously proposed Jacqueline Auriol for the Harmon Trophy, which she was to win then and on two other occasions. Her speed record, the first to be gained by any French pilot since the Second World War, resulted in her acceptance as a *pilot de servitude* at the French Flight Test Centre in Brétigny. A long and arduous period of acquiring qualifications in gliding, instrument-flying and so on followed before her serious test-pilot training began. She qualified in her mid-thirties, working, as she wished, always on completely equal terms with men.

She broke the sound barrier only a few months after Jacqueline Cochran, in a Mystère II, and from 1951 to 1963 she became the fastest woman in the world five times over (with the other Jacqueline soon winning back each of her newly set records). In the later stages of her career, Jacqueline Auriol was to become an aerial ambassador for her country and a highly respected advocate for the place of women in aviation.

Jacqueline Cochran seemed to have a foot – or a wing – in every aspect of female flight from the 1930s to the 1960s. She claims in her biography that she 'became, in effect, the first hostess' in the pre-flight attendant days of her early career, when she looked after the comfort of passengers. It was then, she says, the co-pilot's job to serve the meals, etc. Jacqueline undertook 'all this cabin work' if she could then be allowed 'to do the flying for a few hours . . . I probably put in three hundred hours at the airline controls by this strategy and it was the finest sort of training.'

The air hostess (more recently the 'stewardess' or 'flight attendant') is in a sense the Cinderella of airborne females, although the public has always invested this career with glamour. America was the first country to employ hostesses, with Ellen Church in 1930. Until the beginning of the 1940s, hostesses had also to be qualified nurses. Ellen worked for United Airlines for eighteen months when, after injuring her foot in a car accident, she returned to nursing. During the Second World War she joined the Air Evacuation Service of the Army Nurse Corps, and in September 1944 became the first Flying Nurse to receive the Air Medal.

In 1940 standard requirements for would-be hostesses in the USA were to be between twenty-one and twenty-six years old, between 5 feet 2 inches and 5 feet 6 inches tall, to weigh – proportionate to height – between 100 and 125 pounds, and to be a US citizen, either single or widowed. Each girl had also to be personable, intelligent and attractive. Early photographs of hostesses suggest that they had these qualities in abundance. However, the requirements could be – and often were – used discriminatively against women from black and other minority groups. Airlines, believing that most of their passengers were white, middle-class men, felt that they would be most at ease with white, middle-class hostesses.

The emphasis on the girls' single state was also discriminatory. Hostesses were expected to fade away from the aeronautical scene before they reached their thirties. As one Transcontinental and Western Air (later Trans World Airlines) instructor is reported to have pointed out to his class of trainee attendants: 'If you have not found a man to keep you by the time you're twenty-eight, TWA won't want you either!'

England was slow to employ women as stewardesses. In a 1934 *Girl's Own Paper* Captain W. E. Johns explores girls' possible careers in aviation in an article called 'Looking Up for a Living'. He discusses a letter from a reader who asks how she can become a hostess, and comments, 'I'm afraid this young lady has been reading American air books, or perhaps she has seen *Air Hostess* on the films.' He admits that it is unfortunate that 'we do not have air hostesses in Europe' when 'nearly every passenger plane in America has one'. Friends of his have spoken very highly of these American girls, and he would welcome the introduction of stewardesses on English airlines. As he says, in jocular mood, 'a young and attractive girl casually serving coffee in an airliner, possibly in rough weather, is certain to create a feeling of confidence among the passengers.'

Only three years later, in 1937, the *GOP* again referred to air-hostessing as a career in an article by W. A. Bagley called 'Jobs in the Air: The Air Stewardess'. It is headed with a photograph of an attractive, uniformed, unnamed American stewardess, together with a picture of 'Britain's first air stewardess' who is also, unfortunately, not named. She is, the captain tells us, 'performing one of her many duties. Remember how difficult it is to pour out on a train, therefore what must it be like on an airliner?'

This pioneering British stewardess has no uniform. She wears a dark, polka-dot dress with a very large, ruched, white collar. One cannot resist the feeling, when comparing the two photographs, that the Americans are going for wholesome efficiency while the British strive for a cucumber-sandwiches and tea-on-the-lawn image of graciousness.

Bagley gives his readers lots of facts and figures about American airlines' requirements for their hostesses, but all he can tell us about potential English employers is that it is Air Dispatch Ltd who have employed the polka-dotted lady – 'a former typist – twenty-one years old – to be hostess of their new eighty-minute Croydon–Paris Service'. He warns girls that this is 'definitely a job of the future' and begs them not to inundate airlines with 'applications for jobs which do not, *at present*, exist'.

It was not until the Second World War that the British Overseas Airways Corporation employed stewardesses on its airliners for the first time. British and American girls who did such work during this period were, of course, often exposed to great dangers in the air. Certainly by the time the war had ended, there seemed to be a new respect for them. *Girl* caught the laudatory mood in 1958 with the start of a long-running series written by Betty Roland and drawn by Dudley Pout about 'Angela, Air Hostess; the story of a girl who longed for adventure'. She got it, too. Although in some respects it seemed retrogressive for the paper that had started off in 1951 by starring a team of girl flyers (Kitty Hawke and Co.) later to focus on an air hostess in its sole aviational feature, Angela pulled in youthful readers and retained her popularity over a run from August 1958 to March 1960.

Angela has to overcome maternal opposition to an airborne career. Her gruelling but exciting training with Wingways Airline is vividly described. She is unfailingly trim and efficient, and sometimes concerned about her passengers' welfare even beyond the call of duty. Because her picture-strip stories occupied only one page each week, the scope of her exploits is somewhat limited. Nevertheless she copes with political stowaways from Eastern Europe, nervous passengers over the Atlantic, a kidnapped child in Greece, and so on, with aplomb.

She is generally intrepid, with the courage of her convictions, but on occasions when she becomes infatuated with one of Wingways' pilots or some other personable male, she wilts into temporary wishy-washyness. After two unfulfilled love affairs, it takes an older woman – her mother – to rally her ('You're young, you're free, you've got the whole wide world in front of you!') into setting up a private charter company with two ex-Wingways pilots.

In real life, by the mid-1940s, American air hostesses, realising that they were offering a great deal in terms of service and prestige to their employers, began to demand better conditions and pay. Ada Brown, who started as a stewardess with United Airlines in 1940, led the movement towards unionisation of flight attendants.

Lucille Chase, an American stewardess in the 1950s, described the job as a combination of 'hat check girl, nurse, babysitter, mother, cook, waitress, diplomat, psychiatrist, confidante and companion' and her book, *Skirts Aloft*, referred to the problems of male chauvinism in the cockpit. A decade later, the 1964 Civil Rights Act was a factor in improving the lot of US stewardesses. Although the attitude of many airline executives towards them was still 'Use them till their smiles wear out: then get a new bunch', protection of employment was on the way. Regulations forcing the girls to resign on marriage, or on reaching the (by now raised) age limit of thirty-two to thirty-five (depending on the airline) were fought in court and by unions. (Airlines' discrimination against married women, apparently, had been because a member of a two-income household could afford to be more militant than a single woman when bargaining for better wages and conditions of service.) Although Asian, Native American and Hispanic women had been recruited for a few routes (for example, in areas such as Hawaii and Latin America), their numbers were few, and virtually no black women were employed until after the passing of the 1964 Civil Rights Act. In Britain, too, discrimination has decreased and, as in the USA, female flight attendants are now viewed as career professionals and no longer as dollybirds.

Air hostesses remained popular role-models in girls' fiction for at least three decades beyond the early adventures of *Girl's* Angela in the 1950s. Editors and authors apparently found girl hostesses more acceptable as heroines than female pilots, presumably because they had the more cosy, caring-for-passengers image, and – implicitly – they left the actual piloting of the planes to men.

In 1970 *Bunty* featured Wendy Round-the-World: 'Wendy Brown's love of children has won her a job as Junior Air Hostess to Elmbrook Airlines.' She 'flies all over the world' looking after 'youngsters travelling alone on the company's planes' and, we are told, helping out wherever she is needed. There is an interesting ambivalence about Wendy's age and status – and indeed the age of some of her charges. As a Junior Air Hostess, she is presumably in her late teens, but sometimes the 'youngsters' whom she looks after are about the same age (eight to twelve) as *Bunty*'s readers. It is therefore not clear whether Wendy really is the character in the stories with whom they are supposed to identify. She is enterprising enough, and generally on the side of the kids, but occasionally a bit schoolmistressy with them ('All right, Gretel, let everyone down if you want to . . . I didn't realise you were such a selfish little girl!') – although, always, of course, for their own good.

'Noreen, Air Hostess', in Hamlyn's *Exciting Stories for Girls* (1975), is written by 'Heather Granger', who was really John Wheway, a prolific author of both boys' and girls' stories in the Amalgamated Press's weekly papers from 1919 until the late 1960s. His writing ranged from historical romance to sporting thrills, from dramatic sleuthing mysteries to 'tales of hearth and home', from career adventures to stories with conventional school settings. (For most of the 1930s he produced the weekly tales of Bessie Bunter and the girls of Cliff House School in the popular twopenny paper, *The Schoolgirl*.) Heather Granger was one of his many feminine pseudonyms.

Noreen Farraday, an air hostess with the International Air Corporation, is one of his lower-key heroines. The first paragraph of the story promises her involvement 'in some very exciting moments in the Middle East' – which, in the girls' fiction context, suggests hair's-breadth escapes from power-mad if not actually rapist sheikhs. However, Noreen simply has to get the pilot to land the air liner in Bahrain, an unscheduled stop on their India to London flight, because a small baby in her charge goes off his food and begins to turn blue. With no doctor or nurse on board, Noreen correctly diagnoses 'some sort of heart trouble'. At Bahrain she rushes across the city with the baby in her arms to find a 'famous heart surgeon' who is on holiday there. He gives the child an injection 'to keep him going for twelve hours' but stresses that he must be rushed to London for surgery. Noreen's 'exciting moments in the Middle East' then begin in earnest, as, with time running out for the baby, she tries to return to the airport and has to force her way through a street carnival celebration which involves 'a line of camels stretching from pavement to pavement . . . a procession of carts, cars, oxen, mules, dancing youths and girls that stretched as far as the eye could see' and, of course, fearfully out-of-control crowds.

Predictably she wins through – not only saving the baby's life, but ultimately reuniting his mother with the family from whom she had been estranged. As every reader would have expected, Noreen's air hostess training has stood her in good stead.

CHAPTER TWELVE

'This is Your Captain Speaking . . .'

They were girls who could not sit beside
The hearth and let go by
All the joy and pride and thrills that ride
With rovers of the sky.

LOUIS DE JEAN
'To the Ninety-Nine'

A LTHOUGH the achievements of the two Jacquelines focused attention on both sides of the Atlantic on female flying capacities, it was still to be some time before 'top jobs' became available for women in civil aviation. The girls who had ferried planes from factories to airfields during the Second World War and who were determined to continue to fly afterwards came up against a great deal of prejudice. Equal opportunities in aviation were far from being available and, as the *Smithsonian Study in Air and Space No. 7 (United States Women in Aviation 1940–1985)* reported in 1990, most women in aviation during the 1950s flew only recreationally.

The market, of course, was flooded with male pilots released from the services after the war, and the days of extensive cheap holiday flights, when more pilots would have to be employed, were still to come. It would be a long time before the idea of women piloting airliners became acceptable to the general public; few travellers, apparently, relished the idea of hearing a female voice announcing over the intercom, 'This is your Captain speaking.'

There was – and still is – the problem of pilot training. Most men have entered civil aviation after acquiring flight training and experience in one or other of the armed services. Until very recently no such opportunities existed for women in

Britain; those hoping for jobs as commercial pilots simply had to train at their own expense (in the mid-1980s, the cost of obtaining a commercial pilot's licence was estimated at something in the region of $45,000).

Nancy Bird, in her 1961 autobiography, *Born to Fly*, said that the attitude of the public towards women pilots was still as unprogressive at the time of writing as it had been during her early flying days: 'You can give them an Amy Johnson at the controls and they won't fly with her, but give them any youth with two rings of braid on his sleeve and they feel they're all right.' However, there were some important breakthroughs during the 1960s. In America, women in aviation were at least equipped with the legal tool of the 1964 Civil Rights Act. The fact that this had come into being meant that the concept of equal rights *did* strike a responsive chord in the majority of Americans. More and more women began to assume positions of responsibility in the aircraft industry, as air traffic controllers and senior executives of airlines. Indeed, some women became part-owners of airports and air companies, although, as the *Smithsonian Study* succinctly comments, 'More women are flight instructors than airport owners.'

In Britain women were employed in the air force as fighter controllers during and after the Second World War. Pauline Gower was the first ever woman to be appointed to the board of an airline, joining the then British Overseas Airways Corporation in the 1940s. Almost twenty years later, in 1966, Alison Munro ('who comes to us with a wealth of experience gained in the Ministry of Aircraft Production during the war and recently as Under-Secretary, Overseas Policy Division, Ministry of Aviation') was elected to the British European Airways board. Memories were short, and several people, overlooking Pauline Gower, thought that Alison Munro was 'the first woman to be appointed to the Board of any state-owned airline in the world'.

Three years later, in 1968, Yvonne Sintes of Dan-Air became the first flight captain with a commercial airline. By the beginning of the 1970s, women in America and various European countries were being employed as pilots on some of the smaller airlines, and on 19 July 1984 two American women, Lynn

Rippelmeyer (who had begun her career in aviation as a flight attendant) and Beverly Burns, made history when they captained two different Boeing 747s. Both flights were for PEOPLExpress (later to become part of Continental): Lynn flew transatlantic and Beverly cross-country.

The fashion for making and breaking records had faded in the aftermath of the achievements of Jacqueline Cochran and Jacqueline Auriol. So many routes had already been mastered, and technological advances had to a large extent shifted the emphasis in flying from the individual to the machine. Nevertheless, in 1964 Geraldine ('Jerrie') Mock and Joan Merriam Smith became unintentional rivals in round-the-world flights. These were the first such attempts by women since the disappearance of Amelia Earhart during her ill-fated 1937 trip. Joan Merriam Smith's route was a re-creation of Amelia's: 'I had had the dream for years . . . When I was in high school, I would tell my friends and classmates that someday I was going to fly around the world just like Amelia Earhart. Everyone just laughed.' Joan realised her dream but, ironically, she did not become the first woman to fly around the world. That distinction went to Jerrie Mock, who had applied to, and received official sanction from, the National Aeronautical Association just before Joan had done. Joan made the flight anyway at about the same time as Jerrie. Both were successful, and media attempts to turn their separate attempts into a race had the positive effect of drawing public attention to the participation of women in aviation. Jerrie flew in a Cessna 180 called the *Spirit of Columbus*.

The tall, striking and stylish British pilot Sheila Scott (*née* Sheila Christine Hopkins) was then to make some remarkable contributions to aviational achievement. Between 1965 and 1972 she was in the public eye for her long-distance solo flights, breaking over 100 light-aircraft records and injecting into the jet age a breath of the pioneer spirit of the inter-war years. She only started to fly in her mid-thirties, in an effort to build a new life for herself after an unhappy childhood, an unsatisfactory marriage, several disappointing love affairs and a failed theatrical career.

Her loneliness and sense of insecurity were acute. Flying

seemed to offer the only sustained satisfaction she knew; it was a counterbalance to her constant struggles against mental illness and drug and alcohol addiction. Like Jacqueline Cochran, Amelia Earhart and some other female pilots, Sheila felt that she had extra-sensory powers. Unfortunately these were sometimes of a negative nature. During her participation in the 1969 England to Australia Air Race, for example, the failure of many of her instruments coincided with her hearing strange and hostile voices in and around the cockpit. The worst of these came as she approached Darwin, her destination; a female voice suggested that she should pray before she died. Not surprisingly, she felt forced out of the race, making a landing on an island nearby, unable to sleep for three days and running a temperature. Continuing bouts of depression and fear of persecution often prompted suicidal impulses.

In the context of her fairly constant illnesses and frustrations, her achievements are the more remarkable. After Jerrie Mock's and Joan Merriam Smith's round-the-world flights, Sheila decided to take up their challenge on Europe's behalf. She chose the longest possible route of 32,000 miles round the Equator, flying solo in a light Piper Comanche 260B. Her success made her an international celebrity but did little to increase her sense of security. However, she wrote proudly for her school (the Alice Ottley in Worcestershire) magazine in 1966:

> Never will I forget the fantastic welcome and friendship round the world – often hundreds of people – and always the remark 'How happy we are to see a *British* pilot doing something again.' This made me very proud as always with others I have been passionately fighting the battle to try to save the 'poor relation' British Light Aviation for some time now.

Struggling constantly but not always successfully for adequate sponsorship, Sheila broke many light-aircraft records. One of her most significant achievements was to become the first pilot of either sex to fly solo by light plane (a Piper Aztec D) from Equator to Equator over the North Pole. She broke Amy

Johnson's London to Cape Town record, was the first British woman civilian pilot to fly through the sound barrier, and the first British female helicopter pilot.

She was awarded the OBE in 1968, and achieved a prestigious number of international honours. Sadly, she seemed incapable of maintaining relationships and, before she died from lung cancer in 1988 (at the age of sixty-six) was impecunious and very much alone. A serious motor accident in her mid-fifties had left her almost blind, with head injuries and disfigurement. She was unable to fly again, but, gradually regaining her sight and her good looks, took satisfaction in her past achievements and the knowledge that some of her researches had led to greater understanding of the effects of stress on flyers, as well as helping with the struggle against the pollution of the earth's atmosphere.

Judith Chisholm had been working at Heathrow as one of two women air traffic controllers during the 1960s, and as neither the British Airways flying school at Hamble nor the services would accept girls for pilot training, she acquired her commercial licence privately. In 1980, fifty years on from Amy Johnson's Croydon to Darwin flight, she flew from London to Australia to commemorate it, and went on round the world. In a Cessna Turbo Centurion cabin monoplane she set a new record on the London to Australia lap, and cut Sheila Scott's 1966 time by half for the round-the-world trip. Despite her talent, impressive presence and ability as a spokeswoman for aviation, she found sponsorship almost impossible to obtain. She often felt that she was up against sexual discrimination. A male senior pilot, turning down her application for a job, once infuriated her by explaining that the passengers would be unhappy 'if a little dolly-bird started to fly the aeroplane'. This, several decades on, was as retrogressive as an edict of the 1925 session of the International Civil Aviation Organisation that 'A woman transport pilot would have outraged the public.'

Since the mid-1980s most airlines on both sides of the Atlantic have been willing to employ female pilots, but the problem of acquiring training still remains. As a corollary to this, the *Smithsonian Study in Air and Space* provides some daunting information: 'At the midpoint of the 1980s, the United

States has more women in aviation . . . than does any other country in the world. It is revealing, however, that the ratio of women to men in aviation in the United States has not changed a great deal since 1940.'

Women's battles for acceptance as civil pilots began to filter into popular fiction in the early 1970s. The themes of *She'll Never Get Off the Ground* by Robert J. Gerling (1971) and *Flying High* by Sally Wentworth (1982) are similar, although there are interesting differences in their treatment. The first book has an American setting, and the second an English one.

She'll Never Get Off the Ground starts with Horace Studebaker, the personnel director of Trans-Coastal Airlines, waiting to interview a prospective new pilot, Dudney Devlin, whose credentials seem ideal for the job. All the initial correspondence has been conducted by Studebaker on the assumption that the applicant is male, but of course when the ambivalently named Dudney appears, it is obvious that the new, would-be pilot is female. She has already flown with the small air-cargo company founded by her father (who is now dead), but has a driving ambition to become a passenger-carrying pilot 'on a good airline'.

Studebaker, a kindly, paternal character, finds it hard to fault Dudney's personality or manner. He is moved, too, by her real passion for flying, her honesty and her looks. Hopefully he suggests that he'd very much like to have her 'in the, ah, family, so to speak' and that she might 'start out' as a stewardess. Dudney, of course, refuses, and is sent off to her motel for a few hours while Studebaker consults with a higher authority. Jason Silvanius, vice-president, public relations, thinks at first that she can easily be turned down, but both men have second thoughts about this, fearing that she will run to the Equal Rights Commission and 'nail us to the cross if we reject her'. Studebaker begins to feel that she should be given a chance – 'I've got a strange feeling she'll make us one hell of a pilot' – but Silvanius sticks to his guns: 'Personally I wouldn't have hired Amelia Earhart. Too many problems.'

Tom Berlin, the airline's president, is called in. His objections are more crudely expressed; the idea of a woman piloting a Trans-Coastal plane is 'like a dose of clap on my wedding-night . . . A broad in the cockpit? Hell, I'd get off an airplane myself if I saw some damned skirt sitting on that right seat.' Ultimately, however, despite their macho expostulations and epithets, they realise that, because of the Civil Rights Act, they will have at least to make a show of accepting Dudney. They will find some way of hiring her to avoid the appearance of discrimination, but will ease her out during training: 'You know damned well a flight instructor could wash out Lindbergh if he wanted to.'

Gloomily they realise that the bright and beautiful Dudney is bound to pass her 'physical'. Their last line of defence against petticoat invasion is Captain John Battles, the senior pilot 'who could turn her down for something as nebulous as wrong attitude'. Although described as respecting women 'in a quaint, courtly, chivalrous way roughly comparable to that of a John Wayne movie cowboy', Battles has a well known 'shit-list' consisting, in order of prejudice, of 'women pilots, women doctors, women lawyers, and women airline executives'.

The shit-list rather than the chivalry comes to the fore when Battles interviews Dudney, but she has a reasoned answer to even the most acerbic of his presumptions. Her directness and obvious expertise grind him down to the point where he agrees to take her on – though he threatens that she'll be thrown out instantly if she doesn't toe the line in every way, or if she is even five seconds late for any training sessions.

Studebaker, Silvanius and Berlin have nothing left to do about the situation except to contemplate Dudney's legs and wonder 'if she pulls off the miracle and flies for us' whether these should be covered by slacks, or if Dudney should 'wear a skirt in the cockpit'.

The author makes the training process vivid and gripping: Dudney is one of the most competent trainees but quickly realises, when she corrects a fellow (male) student's technical error, that she'll virtually have to apologise for doing so. A low profile as well as a low status seems necessary. (Readers with feminist convictions may find Dudney's behaviour a touch irritating at times; she does seem to bend over backwards psychologically to understand and placate the prejudiced attitudes which surround her.)

Dudney falls between several stools: the guys in her training group resent her if she forgets to batten down her brain-power; the stewardesses fear that she'll put on airs, and christen her 'Wonderwoman', and some of the senior pilots (flight captains) with whom she eventually works are appallingly rude and anxious to cut her down to size.

At times the cockpit is filled with both spoken and unspoken sexual innuendo, while Dudney is being evaluated as a woman by one or other of her male colleagues. There is also quite a lot of sexual symbology in the way in which the various pilots handle the planes. Dudney remembers her father telling her that a pilot should handle 'an airplane like a lover should handle a young girl. Be gentle, but be persuasive and firm if you have to be.' Her respect for one particular captain, who has treated her like a real s.o.b., is enormously increased when she sees that he flies his plane 'like a skilled seducer coaxing a passionate but balky virgin into submission'.

Dudney makes out, and becomes one of the gang, or rather of the various gangs. She also falls in love with Captain Mark Ashlock, one of the most tolerant and liberal of her colleagues. However, she finds that his liberality gets a bit tarnished round the edges when they talk about marriage. He takes it for granted that she will be resigning from her job, and when she points out that she wants to go on flying he rather pettily condemns her for hoping that one day she'll be promoted to captain. He gives her the old routine about women not having the ability to command, to work with subordinates, etc.

The situation resolves itself in a suspenseful 'drama of the skies' (which, like the ending of a thriller, should not be re-vealed). Suffice it to say that Dudney makes an error (to save a life) and that her mistake seems to be judged more as a woman's than as a pilot's or a human being's weakness. In the end *she* gives the airline a graceful way out of their contract – unsatisfactorily, one feels, as this slots her once again into the after-all-women-really-are-only-suitable-for-back-up-roles situ-ation. Nevertheless the personnel director speaks firmly of 'a *very* happy ending': after her wedding and honeymoon Dudney is to be transferred to work in simulator training, the implication being that this will fulfil her. Of course it also means that, as the book's title suggests, she'll never get off the ground again – at any rate, not in her professional capacity.

Robert Gerling's novel provides fascinating insights into pilot training and the inner workings of a big airline, despite its flawed ending. The heroine of the Mills & Boon romance *Flying High* by Sally Wentworth has, in the end, a rather better deal than Dudney. She is Leigh Bishop (the first thing a female potential pilot needs is a gender-ambivalent name) who is taken on as a pilot by Allerton's Air Charter Co. The appointment seems to have been made entirely by correspondence, so that when she meets her new boss, Bryce Allerton – the owner of the company – he is appalled to find that she is a woman, and accuses her of trickery and basic female hopelessness. However, Leigh comes through a flight test, which he immediately insists upon, with panache. Even so, he is unwilling to have her on his staff, so Leigh reminds him sweetly about the Sex Discrimination Act. He snaps back that she is 'a flying fanatic' – which, of course, he would never have done if she had been a male pilot.

Bryce also basely alters Leigh's contract of employment, to her disadvantage, and subjects her to sexual harassment. She doesn't see it that way, because she begins to fancy him, but it really is awkward for her to have a boss who announces his determination to bed her, and then gives her all the duff, smelly cargo jobs, and bawls her out for her mistakes. 'She cursed herself for not having checked the delivery notes . . . But then she remembered that she had been thoroughly kissed [by the incorrigible Bryce] not two minutes earlier, and wasn't surprised that she hadn't been functioning properly.' (So much for feminine job expertise).

The other pilots, all of whom are male, 'treated her as if she was some sort of freak, definitely not to be taken seriously', although they mellow somewhat as she proves her worth. But Bryce continues to blow hot and cold; Leigh simply never knows where she is with him. She even tries to resign, but he refuses to release her from her contract.

Although the romantic theme is the predominant one, there are some intriguing flying sequences in the story. The most grisly is when Leigh, piloting a solitary male passenger, Mr

Hollander, on a specially commissioned trip, finds things getting out of control: 'Do you know what I'd like to make this day really perfect?' he demanded, his eyes on her face, strangely glittering and intent . . . the hand holding her arm tightened. 'I'd like to join the Mile High Club.'

This club, we are told, has a 'very exclusive membership; only people who'd had sex over a mile up in the air were eligible to join.' Hollander offers Leigh a high price for her favours and, when she refuses, he sets out to rape her. He knows that the machine is on auto-pilot and drags her from her seat. A bloody and desperate struggle takes place, and Leigh just manages to get to the control buttons and take the plane off the auto-pilot. It penetrates even Hollander's drink-crazed brain that they are beginning to go into a near-vertical dive; he reieases Leigh, but once she has righted the aircraft he tries again to drag her from her seat; she cleverly uses her aeronautical know-how to keep him at more than arm's-length: 'Cringing under his grip, Leigh slammed on right yoke and rudder and sent him smashing into the side of the plane as it banked steeply.' When he comes at her again she goes into a steep climb which makes him fall backwards: 'There was no way, now that she had this beautiful weapon in her hands, that he was going to win.' Leigh puts the Piper Aztec through 'some manoeuvres that no self-respecting pilot would ever do with a passenger on board' and she doesn't stop until the stench of vomit alerts her to the fact that, at last, Hollander is in a state of collapse. (As a footnote to this sequence, a woman pilot has informed me that the Mile High Club *does* exist, and that 'there is probably a supersonic club too, by now'. She also mentioned that a friend doing air-taxi flying sometimes found his passengers on the way home from the races 'a little worse for wear', so he used to go on oxygen and climb to 13,000 feet or so, which would put them to sleep: 'The effects of alcohol and altitude did the trick.' Producing turbulence could also 'make them all sit down and shut up'.)

The attempted sexual assault in the plane, and a near-miss with a powered hang-glider which suddenly swings across her flight path, are too much for Leigh. She loses her nerve and begs Bryce to accept her resignation. He refuses – and knows exactly what she has to do. 'He half dragged her out into the open . . . He bundled her into the cockpit and fastened the seat-belt round her, plonked the headset on her head. "Now take her up," he commanded.'

At first she is paralysed with fear, but Bryce's 'Trust me, Leigh' and what she reads in his face give her the strength to take up the plane. Her nerve is restored and, more fortunate than the dedicated but unfortunate Dudney Devlin, she knows that in the future she *will* be able to get off the ground professionally. She will also, presumably, have a large stake in Allerton's Air Charters, as she is going to marry the company's owner.

CHAPTER THIRTEEN

Space- and Super-Girls

The Supercats – girl astronauts from *Spellbound*, 1976.

We may not be discharging the correct chemicals into your atmosphere! We can only hope, Rhona!

'The Supercats Spacecrew',
Spellbound (16 October 1976)

S TRICTLY speaking, the female preceded the male into space. The first living creature from Earth to experience orbital flight was Laika, a bitch, who was carried aloft in Russia's second satellite, *Sputnik 2*, on 3 November 1957. (Their first was rocket-launched a month earlier on 4 October.)

Laika spent seven days in space, and the purpose of her flight was the measuring of readings of cosmic radiation; its effect on her behaviour was to have a bearing on whether, or when, humans would be sent into space. Understandably, in view of the tremendous feats of space technology which have succeeded it, Laika's flight is now rarely referred to. However, Junior Lieutenant Valentina Vladimirovna Tereshkova, who made forty-eight orbits of the earth in seventy hours and forty-one minutes in *Vostok 6* on 16 June 1963, will always be remembered as the first woman to go into space. Twenty-six years old, with some experience of parachuting at her factory flying club, she enhanced interest in and the prestige of the space mission in which she participated. Her flight was at least a signal that space did not have to be an exclusively male preserve. She received the honour of Hero of the Soviet Union, married a cosmonaut and became an enthusiastic proponent of peace and understanding between nations.

In fiction various statutory women went into space long before Valentina. As long ago as the 1920s, British comics had established the tradition of devoting one strip to a female character, although 99 per cent of the pictures and stories would celebrate the exploits of male heroes. When *Eagle* took the comics' and boys' papers' world by storm in 1950, the feature which became a cult with most readers (of both sexes and all ages) was the Dan Dare strip. Daniel McGregor Dare's regular team of space explorers included Albert Fitzwilliam Digby (his batman), Sir Hubert Gascoigne Guest (a distinguished middle-aged astronaut who is 'controller of the interplanet Space Fleet') and Professor Jocelyn Mabel Peabody, who was the brains behind many of the expeditions.

As one of the first women in popular fiction to go into space, Professor Peabody was rather more than the token female whom comics readers might have expected. She was, from the beginning, an impeccable lady who rarely let her hair down, either physically or figuratively. A natural and unaggressive feminist, she is a Doctor of Philosophy who was at one time 'the youngest lecturer at Oxford'. We are told that she was Professor of Plant Biochemistry for an unspecified period at an unnamed university. As well as being a research biologist and an expert on nutrition, agriculture, geology and botany, she is a dab hand at linguistics. She holds a Space Pilot Class 3 certificate, was assigned to Dan Dare's first expedition to Venus by the World Government Organisation, and travelled with him thereafter on most of his major explorations.

Neither Dan nor any of his male colleagues takes liberties with Jocelyn Peabody: despite the intimate confines of space capsules, etc. she is rarely even addressed, or referred to, by her Christian name. It is true that she has, at first, to cope with some fairly predictable anti-feminism from the elderly and conservative Sir Hubert, but her extraordinary competence and ease of manner soon ensure her integral place in the team.

Her scientific know-how often saves the expedition from disaster, and she is not only brainy but plucky and pretty too. (Like most of the original characters in the Dan Dare stories, her appearance was inspired by a real-life person. Frank

Hampson used Greta Tomlinson, a member of his team of artists, as the model for Professor Peabody.)

Her role and characterisation remained stable throughout the first ten years of the *Eagle*'s original run (1950–60), after which she no longer appeared regularly. When the paper was relaunched in March 1982, the Dan Dare strip was a bastardised affair which did not include the authentic characters. It featured an unsatisfying descendant of Dan Dare, and Professor Peabody was conveniently wiped off the space scene by having supposedly committed suicide some time before.

However, when the new *Eagle* appeared in 1989, with Dan Dare drawn by Keith Watson, who had worked with Frank Hampson in the 1950s, the strip's original character and glories were restored, and Professor Peabody fitted as neatly and unobtrusively into the team as she had done during the paper's heyday. Older readers, however, were soon to feel nervous about press publicity which suggested that she and Dan were to enter into a romantic relationship (shock horror!). Dialogue between them became tinged with sexual awareness – 'I certainly wouldn't choose it for a holiday, eh, Jocelyn? Not even a honeymoon!' – and there was media talk of Peabody undergoing a face-lift or 'robotic surgery'. Nothing quite so radical happened, but she was (after forty years) given a new hairstyle; she also abandoned her 1950s' space-age cat suit for 1990s' jogging suits, trainers and (more shock horror) bangles and gipsy earrings. But basically she remained the same good old Peabody, a real sport, far keener on space than sex, and someone on whom Dan's team – as well as readers – could always depend.

Sadly, however, by the middle of 1991 she seemed without explanation to have faded away. From about the time that Britain's first real-life astronaut, Helen Sharman, had got into space, the new *Eagle*'s editors seemed to have decided that females intrepidly hurtling through the ether had little to offer boy readers.

As we have seen, *Eagle*'s feminised companion paper, *Girl* (from 1951), was not sufficiently emboldened by Dan Dare's success to launch any of its heroines into space. Kitty Hawke and her all-girl air crew had to confine themselves to fairly orthodox and basically earthbound detective work. However, as Professor Peabody had obviously inspired enthusiasm, other boys' papers whose heroes were entering the space age also spiced up their resident teams of astronauts by adding the occasional girl or woman. A little while before Valentina Tereshkova's *Vostok 6* orbits took place, the ten-penny Super Detective Library series produced a female psychologist-cum-astronaut, Dr Altha Berins, to help solve a mystery under investigation by Rick Random (chief trouble-shooter for the Interplanetary Bureau of Investigation). The time-setting for Rick's adventures is the twenty-first century. Blond, clean-cut, swashbuckling through space and leaping on and off a bewildering variety of planets, he is also very bright, and strongly aware of his social responsibilities. Protective of the weak, never so much as laying a finger on the curvaceous women who collaborate in his cases, he has only one flaw (and, in the late 1950s, it wasn't considered as one)—he virtually chain-smokes. Even so, he never seems to drop his ash in a messy or inconsiderate manner, although it is not the easiest of tasks to manipulate cigarettes cleanly when one is piloting astro-ships full of complicated gadgets, rocketing oneself solo through space or physically grappling with one's enemies.

Rick's female flyer colleagues are an impressive lot. Dr Altha Berins, for example, is quick to spot in *Sabotage in Space* (Super Detective Library no. 111, 3 September 1957) that 'the madness' which grips thousands of men and women on Terra (AD 2042's name for Earth), making them commit spectacular suicides or murders, is due to the fact that aliens are planting evil thoughts in their minds. She co-pilots the several different space-craft which carry her and Rick on their intergalactic search for the murky mind-manipulators.

It is interesting that Altha, like almost all of Rick's girl helpers, is generally flimsily clad, in what at first looks like a two-piece bathing costume but isn't, because it is made of far richer material and looks fetching with either high-heeled shoes or space-boots, with ultra-modern hats or with astro-helmets. Thus Altha and the female astronauts expose a lot of their comely flesh to the elements, while Rick and his male space explorer colleagues usually sport heavy, high-necked, cossack-style, all-covering garb.

Other space-girl agents or sleuths who grace the series are Lieutenant Brell Canto, a woman pilot who intercepts and captures a UFO, and Marla Orst, daughter of a baron from the Planet Quont which has been occupied by the wicked Ebloni tribe. Marla is as good an astronaut as anyone – handsome, too, although for some obscure reason she has horns growing out of her forehead. She and Lieutenant Brell Canto play prominent parts in *Rick Random and the SOS from Space* (no. 115, 5 November 1957).

Rick Random and the Planet of Terror (no. 123, 4 March 1958) features Feleena Smark, 'special correspondent for the Galactic Magazine and one of the Cat People from the Planet Tigris'. She has long, pixie-type ears, presumably – but puzzlingly – to indicate her feline roots. *Rick Random and the Mystery of the Frozen Plant* (no. 133, 7 August 1958) introduces the female detective, Andi. All of Ricky's helpers are extremely adept space pilots, as indeed they have to be, because they rarely stay long on earth.

Of course their exploits go far beyond Valentina Tereshkova's. Like those of so many fictional girl astronauts, their junketings are not only intergalactic, but involve time-shifts and high-tech paraphernalia so way out that readers have to suspend disbelief and be prepared to topple frequently from space adventures to extremes of fantasy. Nevertheless, if one can accept UFOs which speak with human voices (and, most helpfully, always in English), telepathic 'thought talk', a prolific line-up of different types of aliens, vehicles which function with equal ease in the air, on land and both on and under water, giant dinosaur-like animals, rabbit-eared lions, fish-faced dogs and gorillas with elephantine trunks, the gorgeous girl astronauts who work with Rick Random provide their own brand of excitement and comic relief.

The cross-breeding of space and fantasy fiction encompasses everything from sci-fi to send-ups, with female astronauts cast sometimes as sex objects and victims, frequently as 'vampires', occasionally as partners, and – rarely – as explorers of space in their own right. Often the central girl character is part of a team, or is a skirted version of an already well-established male figure.

Wonderwoman is not exactly an astronaut, although she nips in and out of space (and time) at will. She is one of those super-girls who can traverse the skies or visit the planets without even the help of spacecraft or rockets. Her determination, magnetism or magic hoists her as high as she wishes to go. She and her feisty, fearless, fabulous colleagues – Batman, Superman, Hawkman, Hawk-Girl, Atom, Black Canary, Green Lantern and others – from the nucleus of the JLA (Justice League of America) are a force to be reckoned with. They monitor and benevolently manoeuvre the US of A's public concerns (from Wall Street to military campaigns) as well as the more grisly problems with which some of its private citizens have to grapple.

Wonderwoman's exploits often remind us that, as the publishers state on some of their comic-book covers, 'DC Comics aren't just for kids'. The big-eyed, black-haired, superbly built Wonderwoman – despite the magical powers on which she can draw if required – is not above kicking her male opponents in the balls (at least it seems so from the pictures), or having a sexual flutter with randy mythological characters from antiquity (like Lord Hermes, who has apparently been promoted from messenger of the gods to Olympian god of messengers, and who can assume human form as easily in twentieth-century America as in classical Greece). We are occasionally reminded that Wonderwoman is the Amazon Princess, Diana; that she has been granted the wisdom of Athena, the strength of Hercules, the speed of Mercury and the beauty of Aphrodite. If this were not a sufficiently formidable combination of talents, she also has the sensual appeal of Raquel Welch, and – the only external apparatus she ever needs to supplement her in-built assests – a magic lasso ('the blazing lasso of Hestia, goddess of the hearth, within whose fire no lie can survive, no mere facade can stand').

All this might seem a bit overblown and far removed from

the meticulous scientific approach we expect from our astronauts, but nevertheless Wonderwoman initiates a great deal
of aerospace adventure for us to enjoy, even though it is of
the 'Whoosh', 'Bam', 'Whock' variety. Wonderwoman came
into being a year or two after Superman and Batman, who
were launched in 1939. She can be seen as a feminised
version of both of them, but apparently her creation was
directly inspired by William Moulton Marston, a psychologist
who hoped through Wonderwoman to conteract the bloody,
thunderous and macho tone of the general run of comic
books. In spite of the aggressiveness which Wonderwoman
can employ when events demand, she also often saves and
reforms her enemies.

Her critics, however, have been unimpressed by the supposed softening effect of her romantic impulses. At one
extreme she has been regarded as an embodiment of lesbian-
ism, fetishism and even, on occasions, sadism. At the other,
however, she is respected as an early symbol of feminism.
She was first drawn by Harry Peter for *All Star Comic* (no. 8
December 1941), and was switched in January 1942 to star in
the first and subsequent issues of the *Sensation Comic*.

She has since been produced by several different artists
and writers. Though generally drawn to represent the most
voluptuous of sex objects, she is multi-faceted and always
evolving. The feminist streak that so well suits her
Amazonian roots finds interesting and varied forms of expression. As indicated by the fact that she has voluntarily
'renounced her immortality' in order to give help to others,
she remains basically a nice, all-American girl, living in an
East Side condominium and conscientiously washing out her
star-studded smalls every day.

As the super- and space-girl *par excellence*, Wonderwoman
has inspired many derivatives who seem designed to appeal
to adult American readers. Super-girl and Zatanna slot
firmly into this category; so too does Miss (originally 'Black')
Fury, whose newspaper début on 6 April 1941 predated
Wonderwoman's first appearance. Miss Fury storms colourfully through space, righting wrongs and demolishing her
enemies. She was the first supergirl to be illustrated by a

woman, Tarpé Mills. There is, however, nothing soft or feminine about Miss Fury – 'Get out of here or I'll give you another taste of the lash!' – who blithely uses whips and branding-irons to foil her foes.

For juveniles, British comics have produced their own super-girl (in *Judy* from 1983), drawn by Giorgio Letteri. She is Debbie Danger, who only becomes a flying daredevil when performing on TV as Supergirl. In her private life she is far from intrepid, and is even scared of clockwork mice. Dudley Wynn's Valda (in *Mandy* from 1969 to 1982) not only air-lifts herself frequently but is possessed of extraordinary strength and eternal youth. Vanessa from Venus ('The Most Amazing Friend an Earth Girl ever had'), who arrived in *June* in 1961 in a flying saucer, and remained there until 1967, is dedicated to doing good works throughout the world and, particularly, to helping adolescent heroines who have problems. Dennis M. Reader's Electro-Girl, who flashed in and out of *G-Boy Comics*, *Whizzer Comics* and *Super Dupes* from 1947 to 1949 was possibly the earliest high-flying super-woman in British comics. Her real name was Carol Flane, and she boasted proudly that she 'gave no peace to law-breakers' whom she would 'zap' with terrific bolts of electricity. A touch more subtle, and spin-off from TV and the cinema, Lady Penelope from *Thunderbirds* had a good run in *TV Century 21* (1965–6) and her own comic, *Lady Penelope* (1966–9), while Princess Leia Organa of Alderaan was prominently featured in *Star Wars Weekly* (1978–80) and other comics until 1985.

All these, of course, are larger-than-life characters. Attempts to create fictional female astronauts who are charismatic but convincing and not wildly way out have achieved only short-lived success. Keith Watson's Fran, Sally and Kathy, who made up 'The Space Girls' (*Tina*, 1967 and *Princess Tina*, 1967–8), sounded promising, looked good, but were air hostesses and therefore ultimately dependent on the achievements of male astronauts in their Venus-via-the-Moon-bound space-liner.

It was not until September 1976 that British comics came up with an all-girl space crew, the 'Supercats', illustrated by Badia, in no. 1 of D.C. Thomson's *Spellbound*. This feature ran until the paper ended in January 1978 and was then transferred to *Debbie*. The 'four very special girls who crew Spaceship *Lynx* of Cat Patrol' are Hercula, who has incredible strength, Electra, who can generate electricity from any part of her body, Fauna, who can change colour and shape like a chameleon, and Helen Millar, who is simply described as an Earth Girl, and the captain of the team. The feline mood was figurative rather than actual, with 'cat' presumably used in its 'pop' and not its animal meaning. Helen and Co. are full-bloodedly human and never have cat-like ears, tails or other excrescences foisted on to them, as do some of Rick Random's female colleagues.

The Supercats are big-boobed, lusciously leggy and splendidly built all over, but wholesome rather than sexually alluring. This is appropriate, of course, for *Spellbound*'s readers were almost exclusively female and somewhere in the eight-to eleven-year-old range. They are 'special investigators', and the mysteries which they probe are often of the spooks-in-space variety. They discover hitherto unknown 'deadly and mysterious' planets and save others from 'dying'. They encounter an ancient Spider Queen who enmeshes astronauts in steel webs and then offers them up as tit-bit 'flies' to her hungry spider subjects; they tangle with giant, destructive, triffid-like flowers; with time-shifts ('I reckon we're about a hundred years too late. There's a thick layer of space-fungus on this food'); with great and capital P'd Power ('We're almost a thousand miles from Spiro. To almost hit us from that distance means they have a fairly complex defence system!'), as well as fairly predictable hazards such as ghost space-ships and menacing meteorites, which they take almost nonchalantly in their stride.

In these Supercat episodes, space apparatus and settings as well as the four fetching protagonists are superbly drawn. So too are the robust 'sexy sailors of the Spaceways', Rachel, Sherry, Karry, Carla and Julie, known collectively as 'The Stargrazers', and featured in a series of adventures published in America by Innovation Books in 1989. The back-cover blurbs make it fairly clear that the Stargrazers are expected to appeal to adults (and presumably to men rather than women). They promise 'stirring "good girl art" space adventures'. Those readers who can't get enough of the 'sexy sailors' or 'good girl art' from the books are invited to 'Take a Stargrazer Home' by purchasing a portfolio of pictures or a large poster of the girls. Without exception, these five astronauts are handsome, luxuriantly coiffured, splendidly busted, small-waisted, sumptuously thighed and long-legged.

Incidentally, a definition of stargrazing (as distinct from stargazing) is provided on page 31 of the first story, 'Here There Be Dragons'. Julie, the new recruit, is informed that 'We graze a star and skim off tons of hydrogen – free fuel for carrying cargo across stellar distances'; also that an understanding of science rather than 'magic or sorcery' makes 'stargrazers tick'. Rookie Julie performs an important function in the unfolding of the occasionally complex plots, because the more experienced stargrazers have to enlighten her (and thus indirectly their readers) about a lot of what is going on that is too subtle to come across in the pictures and speech-balloons. Julie, a little less super-girlish than her crew-mates, strongly resembles Marilyn Monroe in face, figure and personality; like her, she appears slightly dim – 'Oh my stars!' – yet is knowing and essentially endearing.

Stargrazers float backwards and forwards in time, as well as all over space. We learn that their space-craft *Crock of Gold* is protected from attackers by a dragon-type aura trapped completely in a medallion which derives from the essence of ancient lost stargrazers. Classical legends are referred to and intriguingly adapted. The sirens who tried to lure Odysseus have been transmogrified into 'the sirens of space – gorgeous hunks of manhood, each travelling on the tail of a comet'. (In the same breath one stargrazer complains to another that 'most men we meet are part of the great unwashed': also that 'recycled star-ship air is so stale – and the sirens smell so good.')

The highspot, however, of the activities of Rachel, Sherry, Karry, Carla and Julie – at least from the male reader's point of view – must be their scantily clad exercising sessions, when buttocks bounce, hair swirls and extremely high-heeled shoes (surely unusual for exercise garb) accentuate the length of the all-girl air crew's legs. Despite the Stargrazers' venturesomeness, it is obvious that they are primarily sex objects.

It seems unlikely that any of the extremely professional real-life women who take part in the USA's space missions will identify intimately with the Stargrazers. The most visible women in the aerospace industry are of course the astronauts Sally Ride, Kathryn Sullivan, Anna Fisher and Margaret Rhea Seddon, as was the late Judith Reznick. However, the number of women working as engineers, scientists and consultants at government agencies such as the Federal Aviation Administration and the National Aeronautics and Space Administration continues to grow. In 1980 21.4 per cent (253,900) of 1,185,000 employees in the field were female. By 1984 the figure had increased to 23.4 per cent (280,600) women out of a total of 1,197,200.

America did not send a woman into space until fifteen years after Valentina Tereshkova's 1963 flight, and the organisers of the space programme have been subjected to a great deal of criticism for lagging so far behind in this. As long ago as in the late 1950s, the USAAF were conducting experiments into the physical and psychological suitability of women for space flights. The extremely experienced Ruth Nichols, who had been flying since 1922, recorded in her *Reminiscences* that she participated in these tests at the Wright-Patterson Air Force Base in Dayton. Geraldine ('Jerrie') Cobb was brought into the test programme in September 1959. At twenty-eight years old, with three world records, 7000 hours of flying time and the Fédération Aéronautique Internationale's Gold Wings of Achievement behind her, she had established herself as a successful executive in the aviation industry. However, she was prepared to abandon that career if she could become the first woman to undergo the astronaut tests for the Mercury space programmes at the Lovelace Foundation in New Mexico. Jerrie successfully completed all three stages of these demanding tests and her results prompted several NASA military officials to propose that the USA should take the opportunity of becoming the first nation to send a woman into space. Her performance also led to a programme of tests for a further twenty women, all of whom were skilled aviators. Twelve of them passed the first two rounds of tests and were due to take the third and final round at Pensacola Naval Air Station in Florida. At the last minute, NASA suddenly refused to sanction this, apparently on the grounds that their completion of the test programme might be taken as approval of female astronauts. Not unnaturally the twelve aviatrices felt anger and frustration, and the ensuing controversy about the roles of women in aviation and in space was to continue for a decade.

Jerrie Cobb with Jane Hart (one of the twelve candidates and the wife of Senator Philip A. Hart) lobbied Congress, the NASA administration and Vice-President Lyndon Johnson for women to be included in the space programme. NASA responded by announcing a Catch-22 situation: in future, astronaut candidates must have experience as jet test-pilots – but, of course, apart from the incorrigible Jackie Cochran, it was virtually impossible at that time for any female flyer to test a jet aircraft. Such jobs with civilian contractors were always given to men, who had generally received the appropriate training when doing military service. Really the *only* way to acquire experience of jet test-piloting was through association with the military, and at that time women were debarred from flying with them.

Valentina Tereshkova's launch into space in 1963 was undoubtedly a propaganda coup for the Soviet Union, as Claire Booth Luce was quick to point out in a *Life* magazine article condemning NASA and claiming that its failure to keep abreast with or in advance of the Russians in this respect 'may yet prove the costliest Cold War blunder'.

Be that as it may, America has since adopted a laudable policy of employing women in its space programme not as tokens, but simply if and when they are the people most suitably qualified for the particular jobs. In 1978 Dr Sally Ride ventured beyond the earth's atmosphere when, with five other women, she formed part of a team of fifteen pilots and twenty mission specialists. In 1984 *Challenger* went into orbit, and in its crew were Dr Ride and Dr Kathryn Sullivan, who was to become the first woman to walk in space. Dr Margaret Rhea Seddon was one of the team of the space-shuttle *Discovery* which was launched in 1985 to salvage *Syncom*, a misfunctioning communications satellite.

The internationally televised *Challenger* take-off and crash of 1986 brought spotlight to bear on the two women of the group of seven people who were killed in the disaster. Judith Reznick was an established astronaut and crew member: few people realised until the facts and figures behind the names of the fatalities emerged that she was, in fact, a woman, because crew lists in advance of the flight had been given only by surnames. The mother and teacher Christa McAuliffe had, however, caught the public's imagination as the first 'ordinary member' of American society to be taken into space. She *was* a symbolic figure, but not because she was a female. She was in *Challenger* to typify and represent contemporary American society. The USA's space programme was temporarily curtailed after the *Challenger* disaster but, as it once again gets under way, there is little doubt that its highly qualified women astronauts will continue to play a significant part.

Meanwhile, journeys into space have been achieved by nationals of an ever-growing list of countries. On 18 May 1991 Britain's first astronaut went in to orbit – and she was a woman, the twenty-seven-year-old Helen Sharman from Sheffield. As Britain has no space programme of its own, Helen was taken up as part of the Anglo–Soviet Juno mission, organised in the USSR. She was the third woman to go into space from Russia; Valentina Tereshkova was of course the first, and early in the 1980s Mrs Svetlana Savitskaya was the second.

A former food technologist for Mars, the confectionery firm, Helen was selected from nearly 13,000 applicants. She was naturally asked many questions during her two-year training about her role as a female in space, and came up with appropriate answers: for example, 'The fluid shift in weightlessness might be different because I have breasts.'

Learning Russian, the language of the mission, was apparently one of her most difficult hurdles. There is encouragement to be gained from the coming-together of cultures that Helen's flight represented, and from the fact that a woman was chosen not as a token but on her merits. Her professionalism was much praised by her colleagues, and she played a key role when the Soyuz TM-12 spacecraft was docked manually at the Mir Space Station.

The sting in the space saga's tail, of course, was that the male commander of the mission, Anatoli Artsebarsky, made some chauvinistic remarks which seemed to cast a slur on the young woman who did her job with such admirable determination and efficiency. First he told the *Moskovsky Komsomolets* magazine that space flight 'was not a feminine thing' and that 'with a man, you can get through a big volume of work.' Next he commented in a radio transmission made during the docking at Mir, 'I just think it is hard work, not a woman's work.' Natasha Artsebarsky sprang to her husband's defence, saying that he was a liberal with a sense of sexual equality who would occasionally do the washing-up, and that he did not object to her being a parachutist. British members of the team criticised the Soviet cosmonaut commander's comments, bemoaning the fact that 'many military men are male chauvinists'. Helen serenely denied encountering any prejudice on the trip, which she seemed to have found not only exciting but deeply satisfying: 'I did not want to come back. I was very busy. I could easily have spent another two weeks up in space.'

Apparently she was not the only woman out there during May 1991. While she was in the Soviet Mir Station, the American shuttle *Columbia* sent up three female astronauts as part of a team of seven. They were accompanied by more than 2000 jellyfish, thirty rats and thirty mice on their nine-day mission to study the effects of weightlessness on animal life: 'It's a dream mission for me,' said the 'astronaut doctor', Margaret Rhea Seddon, and it seems that thirty-five years after that engaging Russian bitch Laika blazed the trail in *Sputnik* 2 in 1957, several women *are* having their space dreams and fantasies fulfilled.

CHAPTER FOURTEEN

Per Ardua
ad Aequalitatem

Oh yes, she will go far. And we know where she is going.
But what do we know of the terrors of the journey?

T.S. ELIOT,
The Cocktail Party (1951)

I N 1991, eighty years after Mrs Hilda Hewlett became the
first British woman to gain a pilot's licence, one of her
countrywomen went into space and another got her RAF
wings. As Amelia Earhart had forecast fifty years earlier,
women had indeed got 'some-place' by their own efforts – and
that place was the world of equal opportunity, equal responsibility and equal risk.

Possibly because the last decade of any century is a time for
looking back to make assessments, as well as for looking forward with aspiration, a small but steady stream of retrospective
novels about female flyers has now begun to appear. Harriet
Hudson's *The Sun in Glory* (1991) brilliantly re-creates the early
days of powered flight in Britain, and the contribution to this
of one woman, Rosie Potts. In her preamble to the story the
author is at pains to point out that it was really 'Mrs Maurice
Hewlett in August 1911' who blazed the aerial trail, and that
Rosie is only fictitious. Nevertheless she, her golden machines,
her failures and successes spring off the page with an authenticity that is enhanced by the fact that several real-life luminaries of early aviation come into the story.

Rosie is no conscious feminist, but because she has a flair for
engineering and a passion for flight she turns down the conventional, domestic role that Edwardian society tries to nudge
her into. At the beginning of the book she is an orphan who,
after her Cockney parents die in a train accident, is adopted by
the eccentric industrialist and would-be flyer, William Potts. At

six years old, Rosie remains disoriented by her traumatically sudden change of parentage and switch of homes from the claustrophobic confines of an East End tenement to the expansive splendours of William's Kentish country home, Brynbourne Place. Pugnacious, rebellious and unhappy, she feels no interest in or bond with her new family. One day William forces the bored, defiant child into flying a kite with him, and makes her run with the string:

> Rosie obeyed, running, running, spinning out the cord, feeling the wind grasp it and take it for its own. . . . There it was now, fluttering, dancing, and she had done it; she was part of it. . . . It was alive, this magic thing of wood and paper, and wanting to be free . . . she looked up to see it soaring and sweeping, making faces down at her, calling to her. 'Come up, Rosie,' it was saying, 'come up here. It's so easy.'

It is far from easy, of course, but from that moment Rosie, like William, is hooked on flying. She shares every one of the triumphs and tribulations involved in the building of his first plane, *Pegasus*, eventually becoming his mechanic and, unexpectedly, his first successful pilot. Jake Smith, who has worked with the Wright brothers, is brought into the venture in its early stages, and helps considerably to get *Pegasus* airborne. His life and Rosie's become inextricably intertwined as the glorious, pioneering days move on into the anguished years of the First World War when, of course, a new and horrific use is found for Rosie's precious machines . . .

Another recent book which captures the atmosphere of an earlier period and involves women in flight, is *Some Women Dream* by Helen Mansfield (1990). The Second World War has just finished, and nineteen-year-old Louise meets Joe, who is in England with the USAAF. Having flown a Thunderbolt in battle, he now flies a Tiger Moth for pleasure and shows it proudly to Louise. She goes into 'a tail spin of excitement' when he comes to see her again, and is soon in love – with him, and with the idea of flying. She accepts his proposal of marriage on condition that he teaches her to fly and, before the wedding, she becomes a qualified pilot. Flight and its imagery spill over into every aspect of their lives, even their lovemaking:

'You start by climbing aboard the aircraft.' Louise climbed astride his body. 'Then you manipulate the joystick to make the plane go up and up and up. Jesus, Louise, you want to give me a heart attack! And the rudder to get your sense of direction.' They rolled around the bed in mock flight, searching rudders and joysticks and starters and doing loops and spirals of a new dimension, until their bodies hastened towards the climax of sensuality.

Louise moves with Joe to the USA where with 'more guts than a hog' she joins him in aerobatic exhibitions. They achieve a 'total union of heart, soul and will to succeed' but, sadly, Joe dies when his plane crashes, and his loss prompts Louise to suicide, which she achieves in a spectacular dive at the ending of an aerobatic display.

An interesting development in recent flying fiction is that contemporary heroines are now no longer content just to be pilots: they end up owning their own airlines too. Diana Stainforth's *Friends and Other Enemies* (1989) is an exploration of dualities in relationships – of love/hate, not only between male and female, but between friends of the same sex. Ryder starts out as a spoilt, poor little rich girl who finds it hard to make friends. Her fortunes decline dramatically (affluent, unscrupulous Daddy is imprisoned for financial malpractice, and ruined; her mother commits suicide; her fiancé betrays her; she has to bring up her small brother—*and* she is homeless). Nevertheless she gamely hoists herself by the boot-straps, faces her big-sisterly responsibilities and sets up house in all that remains of her father's estate – a derelict airfield. Gradually she brings this back to life and lustre. Eventually obtaining a pilot's licence, she becomes a high-flyer in more senses than one, building up airlines, air corporations and simply masses of money. She also acquires a passionate polo-playing lover (who lets her down) and an aviator husband (lots more love/hate here) with whom she rises to new physical and emotional heights. Ryder typifies a new trend in high-flying heroines.

In the 1990s, women from wide-ranging backgrounds and age groups are demonstrating, as keenly as the early pioneers, their insistence on being part of the world of aviation – and no longer only on its fringes. As Judith Chisholm has remarked, 'All it takes is determination, an independent spirit and a thick skin.' Evidence of these qualities in abundance is vividly apparent in the 1991/2 *Gazette* of the British Women Pilots' Association. The cover picture shows its president, Naomi Christie, 'walking on air . . . Having slipped the surly bonds of earth for a quarter of a century, Naomi celebrated her seventy-ninth birthday by joining the Association's wing-walking day at Cranfield Aerodrome. . . . She went on to round the day off by doing a few circuits in a Chipmunk.' (Before her official retirement, Naomi spent twenty years as an instructor with the RAF's Gliding and Soaring Association.) As well as celebrating the achievements of vintage flyers, the *Gazette* focuses on the efforts of young, would-be and newly qualified aviatrices. The Association's work in providing practical flying opportunities, promoting training sponsorship and arranging career symposiums is invaluable. It was able to report with satisfaction that at the beginning of the 1990s: 'All the major British airlines now employ women aircrew; others are working as ferry, charter and helicopter pilots, Chief Flying Instructors, Qualified Flying Instructors, Assistant Flying Instructors, glider microlight and balloon instructors.'

Unreasoned prejudice against women flyers still finds voice from time to time, however. A letter in November 1984 in the magazine *Pilot* is a depressing example: 'Cannot the ladies leave entirely to the gentlemen one or two "worlds" – and I plead for aviation for one?' Hopefully the enthusiastic involvement in aviation of such a popular public figure as the Duchess of York may make some of the doubters take a new look at the aspirations and achievements of female flyers. 'Fergie' gained her private 'fixed-wing' pilot's licence in February 1987, after a course of lessons given as a wedding present by the Oxford Air Training School. She acquired her helicopter licence in December of the same year, and has written two children's books with 'whirly' themes. Both of these, *Budgie the Little Helicopter* and *Budgie at Bendick's Point* were published in 1989 and there are more to come.

In 1986 a woman played a major part in what was possibly the most outstanding civil aviation exploit of the second half of the twentieth century – a flight all round the world without stopping. Jeana Yeager shared the piloting with Dick Rutan, and their strange-looking aeroplane, *Voyager*, was designed by Dick's brother, Burt. It was just about the lightest ever built, on a size/weight ratio; a flying fuel tank, in a sense, but a vehicle not only for adventure and endurance but vision.

Jeana Yeager had trained to be the first woman in space in a private rocket launch project which was ultimately abandoned. Meeting Dick Rutan at a Californian air show led to the 'wild idea' of an aeronautical exploit which, despite years of gruelling preparation, was ultimately to test both pilots to the uttermost. They took off from Edwards Air Force Base in California's Mojave Desert on 15 December 1986, convinced that a 'non-refuelled, around-the-world flight was the last milestone in aviation and that whoever did it would be entering into history'. They completed the trip in nine days, spending these in a tiny unpressurised cockpit, and flying through some of the worst weather conditions in the world. Their plane weighed less than a car and was filled to the wingtips with 200 gallons of fuel. (One non-stop world-circling flight had taken place previously in 1949 when 'a US Air Force demonstration aimed at impressing the Soviets with American air power' had been arranged. However, *Lucky Lady*, a B-50, was, unlike *Voyager*, refuelled in the air.)

Jeana's and Dick's flight, and the struggles involved in the designing, building and testing of *Voyager* for which funds had to be raised privately as no government support was forthcoming, are graphically described in the book *Voyager: The Flying Adventure of a Lifetime* (1987). This is written by the two pilots, with Phil Patton. Jeana was first attracted to flying because of a fascination with helicopters which, apparently, reminded her of the dragonflies she had watched as a child 'hovering and manoeuvring so lightly in the air'. She had wanted to fly helicopters but, instead, put the money intended for her lessons into the *Voyager* project.

After several years of preparation, when the machine finally took to the air, Jeana and Dick were told that they 'already had one record: for the longest take-off ever from Edwards Air Force Base. To get off the ground, we had used up all but a thousand feet of the longest runway in the world.' The tiny space of the cockpit presented extraordinary difficulties; apart from the problems of immobility, and desperately cramped positions, there were the hassles of food and drink (carefully pre-packed in light-weight plastic) possibly spilling: 'A liquid spill could ruin a piece of electronics; a loose object could catch and slip a control cable off its pulleys.' Jeana explains in the book that 'going to the bathroom' involved 'fecal containment bags', similar to those used in hospitals and the Gemini space programme. These had adhesive sections so that they could be stuck to the buttocks and then removed. In practice flights, Jeana and Dick put ripe bananas in the bags and dropped these out from a small door in the bottom of the cockpit, successfully clearing the rear propeller. However, when they 'tried it with the real thing' – and the addition of toilet paper – the bags 'would barely fit through the prop door' and were also so lightweight that there was a danger of them being blown back on to the propeller: 'The stuff came awfully close to hitting the fan.' So on the actual flight they had to endure the unpleasant process of stowing the waste on board in a compartment in the right wing reserved for this purpose.

They had an oversized rubber band for exercising their leg muscles, but that too presented some problems for Jeana: 'It was pretty nice until it slipped off my toes, nearly smacking me in the eye. I decided that our physical condition was more likely to suffer than improve using the band, so I put it away and never pulled it out again'. Despite sharing the piloting, and therefore being able sometimes to sleep or catnap, fatigue was an almost overwhelming problem towards the end of the nine days in the cockpit, with foggy, half-hallucinatory spells occurring. It is no wonder that, when questioned afterwards about 'What's it like up there?', Dick answered, 'It's as dark as the inside of a cow.'

Jeana and Dick have often been asked since whether the trip was worth all the effort involved – 'the travail and heartache . . . and the distance it put between us and the other paths of life we didn't take . . . and the fear and the reduction of life to a hangar and then a cockpit-sized space'. Their response is that even if they had failed, or lost the *Voyager* and their own lives, it would have been worthwhile to undertake the project, for the technological lessons learnt, and, more importantly, for 'the human lessons of dreaming and work and determination'.

The acceptance of women service personnel as fully fledged military pilots has been slow in America and even slower in Europe. Bound up with this is conflict throughout society, and often sharply in the feminist movement, between concepts of pacifism and militarism. It is, of course, unreasonable for women to demand the same opportunities as men if they are not prepared to accept similar responsibilities and risks. In the run-up to the confrontation between the Allied forces and those of Saddam Hussein in the 1991 Persian Gulf War, there was a great deal of speculation about the role which American servicewomen would, or would not, play in it. Female nurses, ambulance drivers and other essential backers-up of combatant soldiers, sailors and airmen have, of course, been operating in front-line situations for decades. The war in Vietnam saw the involvement of air force women on active duty, particularly in intelligence and nursing work. 'Many women officers could not understand why it was acceptable for *nurses* to serve in the midst of the battle action, yet unacceptable for women in any other function,' reports the *Smithsonian Study in Air and Space, No.7.*

In August 1967, an article by Bruce Callander appeared in the US *Air Force Times*, entitled 'Why Can't a Woman Be a Military Pilot?' This questioned why the services did not consider training women as pilots of military planes for 'limited, non-combatant duty'; it also suggested that 'male pilots should listen closely for an unfamiliar sound in the cockpit. This could be a feminine knuckle knocking ever so gently on the cabin door.' In the event, the knocking must have been stronger in the American navy, because in January 1973 it was announced that eight women had been selected to enter the service's flight-training programme at Pensacola. Eighteen months later, in 1974, six of these candidates had earned their Navy Wings of Gold. The US army began to include women in flight-training programmes very soon afterwards. There were, however, still restrictions on women in direct combat. It was partly because of this that the American air force continued to lag behind the other two services; they claimed that all their pilots were combat pilots, or at least potential ones, even if they spent most of their careers as instructors or transport flyers. Nevertheless, by 1976 the air force had initiated its first 'test program' for female pilots, and regular training programmes were to follow. There was of course a great deal of pressure on the earliest trainees, who felt that if they were unsuccessful the door might be closed on future women would-be pilots. Captain Kathy La Sauce commented succinctly: 'When you step out of the airplane and take off your helmet, the transient maintenance guy almost falls over backwards. That sort of keeps us going.'

In both Britain and France, even in the early 1980s, arguments were still being put out about the delicacy of women's internal organs making them unsuitable for high-speed flight. (It seemed that potential female pilots must have possessed far more fragile innards than airline stewardesses or passengers, to whom the same argument was not applied). However, in the Netherlands women became qualified in the services to fly F-16s in 1987, and in Canada in 1988 training on the CF-18 began for women jet-fighter pilots. By 1991 the French air force had seventeen female pilots, three of whom were flying helicopters.

In 1989 the Royal Air Force announced that it would accept WRAF personnel to train as pilots and navigators, under the same terms and conditions as men. At last the RAF's famous motto *Per Ardua ad Astra* (Through Toil to the Stars) would have as much meaning for its women as for its men.

In May 1990 the names of the first two female trainees to fly solo on jet aircraft were announced, and in January 1991 the Royal Navy followed the RAF's lead by calling for women to volunteer for flying training with the Fleet Air Arm. Girl navigators were also being trained for both services, and the first of these to qualify was twenty-two-year-old Flying Officer Anne-Marie Dawe. She was interviewed for *Women with Wings* by the airminded writer Jennifer Schofield, a former independent-girls'-school headmistress, who has for some years worked with the RAF at the Officers and Aircrew Selection Centre at Biggin Hill as an academic adviser on cadetship boards. (Jennifer is better known to the public as

'Piers Williams', the co-author with Peter Beresford Ellis of the 1981 biography of Captain W.E. Johns, *By Jove, Biggles!*)

Jennifer writes: 'Anne-Marie's achievement is a landmark for all women who dream of a career in the air, but for her it was just another step forward . . . . Her story is one of determination and high goals, but there is no romantic background of a little girl who always wanted to fly . . . her only experience in an aeroplane before she joined the service was a trip in a commercial airliner.' As a child 'She was a real tomboy, with two younger brothers, keen to join in all kinds of sport.' At Drury Falls Comprehensive School she was in teams for swimming, hockey, rounders and netball, but perhaps her greatest achievement in this area was to become Borough Captain of the Ladies Five a Side Football team. Another feat was to become Head Boy of her school – there happened to be no lads in the Upper Sixth Form in her year, and so Anne-Marie organised the male prefects and the joint Head Girl organised the females.

'It was the order and discipline of the St John's Ambulance Brigade' which she joined during her school-days, together with her visits to air shows, that led her to consider a career in the services. With twelve O-levels and an A-level, she felt that 'the WRAF seemed to offer the best options, but even when she joined in December 1986 she did not think about flying . . . . The decisive moment came in July 1989, when it was announced that anyone could apply for pilot or navigator training, and Anne-Marie applied for aircrew . . . . Her aptitude proved better for navigator than pilot.'

Anne-Marie did her aircrew training in Lockheed Hercules transports at RAF Lyneham, and went on to the Royal Aero-Space Establishment at Farnborough, 'flying in a "two crew environment" transporting passengers. She could take over the controls and make an emergency landing, but hopes she won't have to. She says of her work: "It's quite good fun!"

After their meeting, Jennifer Schofield said, 'It cannot always be easy for Anne-Marie – apart from some Wren officers she is the only woman officer at Farnborough, and her status is so new that when I asked for Flying Officer Dawe at the gate, I was told that someone would take me to *him*!'

A little later in 1991, press releases from RAF Finningley in Yorkshire announced that 'Flight Lieutenant Julie Ann Gibson' would make history on Friday, 14 June when she graduated as the first female pilot in the seventy-three-year history of the Royal Air Force' (see Appendix referring to women with wings in the 1940s and 1950s). After logging 200 hours' training over eighteen months, she was awarded the coveted RAF wings by Air Marshal John Thomson. It was a great moment, not only for twenty-nine-year-old Julie but for all women with an interest in aviation. The knowledge, too, that several further female pilots were in training suggested that Amy Johnson's wistfully expressed 1930s' hope for aviation – 'I do not want it to be unusual that women should do things' – was at last being fulfilled.

In fact, Julie Gibson told me that she did not think too much about being the first in the field. She was simply doing the same thing as the men training with her – 'something that thousands of men have done over the years'. Her father is a retired Royal Navy Lieutenant Commander and her mother a one-time Wren, but from childhood Julie wanted to fly: 'I remember looking up at the aeroplanes when I was little, and thinking I can do that.' She read a lot about early women flyers, and particularly admired the Edwardian parachutist, Dolly Shepherd, and the exploits of the ferry pilots of the ATA during the Second World War.

Julie's schooling began in Malaya, where her father was posted, and was continued in Devon, Middlesex and Scotland. Despite interruptions to her education because of family removals and postings, she acquired an impressive crop of O- and A-levels and, in 1980, entered the City University in London, graduating three years later with a B.Sc. Honours degree in aeronautical engineering. While still at the university she received through Esso an Air League Flying scholarship which gave her fifteen hours in the air and took her to solo standard. She joined the university air squadron as an engineering cadet, and was accepted at the RAF College, Cranwell in 1984. After her initial officer training and aerosystems engineering training, she was appointed the Officer Commanding, General Engineering Flight at the RAF Fighter

Base, Wattisham, in Suffolk. Put in charge of seventy-five men, she comments that she was never aware of any prejudice or resentment from them. Her next job was as an engineering officer to one of the McDonnell Phantom (F4J) Squadrons at Wattisham, where she was in charge of 160 men. She became a weapon specialist and spent two years looking after bombs and missiles used by the tactical weapons unit.

With the Phantom Squadron she experienced the exhilaration of being flown at almost twice the speed of sound in twenty tons of fighter aircraft, and, as soon as the opportunity presented itself, she applied for pilot training. She appreciated the encouragement of the men who worked with her then and previously, but ignored media hype and other distractions as much as possible to concentrate on the job in hand. She told me soon after receiving her wings that she hadn't yet acquired her civilian private pilot's licence, although she hoped to do so soon.

Julie's basic flying training was in the single-jet-engine Jet Provost at RAF Linton-on-Ouse, in Yorkshire. She went on to the Advanced Flying Training Squadron of the Multi-Engined Training Wing at Finningley, and, after becoming a qualified operational pilot, was posted to RAF Northolt, to fly Andovers carrying VIP passengers. It was then the RAF's policy that women should not be involved directly in combat, but both Julie Gibson and Anne-Marie Dawe, when asked for their responses to this, implied that they would be prepared for full involvement, if the need arose.* Both have signed up for sixteen-year periods with the service; they are seriously career-minded and prepared to accept equality at every level.

Women's progress in aviation during the last decade has been enormous, and probably by the beginning of the twenty-first century society will be so used to their achievements that these will pass unremarked. In 1992, however, with effort and strife as well as success still very much in our memories, a quotation from Beryl Escott's *Women in Air Force Blue* remains

* In December 1991, Archie Hamilton, the Armed Forces minister, announced that restrictions on female pilots and navigators becoming combatants would be lifted, although it would be several years 'before any of them can be expected to go into battle'.

appropriate. She passes on the comment of the three-year-old son of a former member of the Women's Royal Air Force: 'My grandpa was a sailor, my daddy was a soldier and my mummy was an air-sick woman!'

FOR NINE decades women from many countries and from extremely varied backgrounds have taken to the air. Much has been achieved, but some prejudices and an overall inequality of flying opportunities for men and women remain. Some of the strictures encountered by Edwardian balloonists and parachutists still crop up in the century's final decade: compared with men, very few women have managed to acquire commercial pilot's licences (the figures in Britain in March 1991 were 14,559 for men and 267 for women), so that although several airlines now employ female pilots, their numbers are extremely limited. So too are training facilities for women at advanced levels. Many male pilots receive training in the RAF which they can put to use afterwards in civil aviation careers, but, until very recently, women pilots could qualify for commercial licences only by spending large sums (at present something in the region of £40,000) for private training.

However, although progress has been far from linear, a comparatively small but immensely dedicated and determined number of women have achieved personal flying triumphs and pushed out the frontiers of aviation generally. In the process, they have by turns shocked, delighted, stunned, thrilled and inspired the international public.

The situation of the 'Big Four' of the 1930s – Amelia Earhart, Amy Johnson, Jean Batten and Beryl Markham – provides food for thought about achievement and motivation. Despite their remarkable exploits, it was no easy matter for any of them to establish themselves in rewarding, long-term careers in aviation. They were operating before the organised feminist movement had begun to have its effect on society, although individual feminist voices had of course been finding expression since the beginning of the century. Sometimes these celebrated flyers were driven not only by ambition but by frustration in their emotional lives. Amy

Johnson took her first flying lessons as an act of bravado after a disastrous love affair and a series of unsatisfactory jobs (so, too, did Sheila Scott in the 1960s): Amy, Beryl Markham and Jean Batten all seemed unable to sustain either fulfilling and lasting relationships with men or abiding friendships with members of their own sex. Of this famous four, only Amelia Earhart seems to have successfully integrated her private and public life. As well as being an extremely skilled aviatrix, she found satisfaction in marriage, and was a lively and articulately practising feminist. She was also better able to cope than the others with the truly enormous pressures of fame and publicity to which they were constantly subjected. Those who followed in the trail which they had blazed were often better equipped emotionally to combine career success with the demands and fulfilments of their private lives.

The effect of feminism on aviatrices and aviation – in both fact and fiction – is difficult to assess. As in so many areas of life, its influence has waxed and waned, but there is no doubt that it is one factor which has helped to imbue many of today's flyers with confidence. Its overall impact on society had created a more expansive atmosphere for women flyers, but – as always – success has come only through dogged, *individual* determination.

As we have seen, fiction has sometimes reflected, occasionally challenged and frequently glorified the aspirations and achievements of the real-life aviatrices. It is an intriguing mirror of society's ambivalent attitudes towards them. Again, it is not easy to assess the complex influences on it of feminism. Progressive and retrogressive moods still jockey for dominance in fictional representations. On the one hand, aviatrices and female astronauts in some of the comic-strips designed for male audiences seem stuck in the sex-object roles which they have occupied for several decades; on the other, they are portrayed in a variety of fiction – hopefully read by both sexes – with the intelligence, integrity and intrepidness which they possess in full measure, and which are not, of course, sexually determined.

POSTSCRIPT

EVEN since *Women with Wings* appeared in Britain last year women have added to their aeronautical achievements. On both sides of the Atlantic they are training as combat pilots and, in space, they are extending the discoveries and scientific research which began just over thirty years ago.

America is now the undisputed leader in space exploration: NASA's programme continues to thrill the world, and the list of female astronauts, technicians and administrators who have participated in it grows ever more impressive. Tribute should be paid to them all, with particular mention of Dr Sally Ride, who ventured beyond the earth's atmosphere in 1978, Dr Kathryn Sullivan, the first woman to walk in space in 1984 and Dr Judith Resnik, a member of the sadly ill-fated 1986 Challenger crew. In 1992 Dr Mae C. Jemison, the first black woman astronaut, came to prominence. She carries on the aeronautical traditions of Bessie Coleman, the world's first black female pilot (see Chapter Two). Coleman, who flew in the early 1920s, felt that the world of aviation was 'free from prejudices' and hoped to bring other black people into it: she would have viewed Mae Jemison's achievements with pride and delight.

It is always intriguing to look back at the work and words of the pioneers and to link these with present day attainments. L. Frank Baum, who as the author of *The Wizard of Oz* knew a great deal about intrepid journeys into different dimensions, wrote as long ago as 1911 a novel called *The Flying Girl*. Published in Chicago under the pseudonym of 'Edith Van Dyne', it features Orissa Kane, an extremely engaging heroine. In the book's introduction, Baum writes of his indebtedness 'during the preparation of this manuscript' to those aviational 'greats' Wilbur Wright and Glenn H. Curtis who shared his expectations that the discoveries and inventions of the future would very quickly surpass all 'that have gone before'. He feels sure that the dazzling fictional feats of the flying Orissa 'may be emulated' in real life even before his book 'is out of press' because 'the American girl. . .already recognizes her competence to operate successfully any aircraft that a man can manage'. He declares: 'in America are thousands of girls ambitious to become aviators'.

We can take pleasure in the fact that so many of them, over the decades, have succeeded.

MARY CADOGAN
JANUARY 1993

BIBLIOGRAPHY

(All the titles listed below are published in the UK unless another place of publication is indicated.)

WORKS OF REFERENCE

Auriol, Jacqueline, *I Live to Fly*, (Paris); (in English) Michael Joseph, 1970

Babington Smith, Constance, *Amy Johnson*, Collins, 1967

Bacon, Gertrude, *Memories of Land and Sky*, Methuen, 1938
The Record of an Aeronaut, John Long, 1907

Batten, Jean, *My Life*, Harrap, 1938

Bedford, John, Duke of, *The Flying Duchess*, Macdonald, 1968

Bird, Nancy, *Born to Fly*, Angus & Robertson, 1962

Boyne, Walter J., *The Smithsonian Book of Flight for Young People*, Atheneum (New York), 1988

Cochran, Jacqueline, *The Stars at Noon*, Little, Brown (Boston), 1954

Curtis, Lettice, *The Forgotten Pilots*, Foulis, 1971

Du Cros, Rosemary, *ATA Girl*, Muller, 1983

Duke, Neville and Lanchbery, Edward (eds), *The Crowded Sky*, Cassell, 1959

Douglas, Deborah G., *United States Women in Aviation 1940–1985*, Smithsonian Studies in Air and Space (Washington, DC), 1990

Earhart, Amelia, *The Fun of It*, Brewer, Warren & Putnam (New York), 1932

Escott, Beryl, Squadron Leader, *Women in Air Force Blue*, Patrick Stephens, 1989

Gifford, Denis, *The International Books of Comics*, Hamlyn, 1984
Victorian Comics, Allen & Unwin, 1976

Gower, Pauline, *Women with Wings*, John Long, 1938

Greene, Graham, 'The Unknown War' (1940), in *Collected Essays*, Bodley Head, 1969

Gwynn-Jones, Terry, *Aviation's Magnificent Gamblers*, Lansdowne Press (Australia), 1981

Heath, Lady and Murray, Stella Wolfe, *Women and Flying*, John Long, 1929

King, Alison, *Golden Wings*, Pearson, 1956

Lindbergh, Anne Morrow, *Gift from the Sea*, Chatto & Windus, 1972
Hour of Gold, Hour of Lead: diaries and letters 1929–32, Chatto & Windus, 1973
Locked Rooms and Open Doors; diaries and letters 1933–5, Chatto & Windus, 1974

Lomax, Judy, *Women of the Air*, John Murray, 1986
Hanna Reitsch, John Murray, 1988
Sheila Scott, Century Hutchinson, 1990

MacGregor-Hastie, Roy, *Flying Machines*, New English Library, 1965

Mackersey, Ian, *Jean Batten: the Garbo of the Skies*, Macdonald, 1990

Markham, Beryl, *West with the Night*, Virago, 1984

Moolman, Valerie, *Women Aloft*, Time Life Books (New York), 1981

Myles, Bruce, *Night Witches*, Academy (Chicago), 1981

Oakes, Claudia M., *United States Women in Aviation through World War One*, Smithsonian Studies in Air and Space (Washington, DC), 1978

Penrose, Harald, *Wings Across the World: an illustrated history of British Airways*, Cassell, 1980

Perry, George and Aldrige, Alan, *Penguin Book of Comics*, Penguin, 1967

Reitsch, Hanna, *Fliegen mein Leben*, Deutschen Verlagsanstalt (Germany), 1951; (in English) *The Sky My Kingdom*, Bodley Head, 1955

Rhodes-Moorhouse, Linda, *Kaleidoscope*, Arthur Barker, 1960

Rutan, Dick and Yeager, Jeana (with Phil Patton), *Voyager: the Flying Adventure of a Lifetime*, Heinemann, 1987

Scott, Sheila, *On Top of the World*, Hodder & Stoughton, 1973

Shepherd, Dolly (with Peter Hearn, in collaboration with Molly Sedgwick), *When the 'Chute Went Up*, Robert Hale, 1984

Spender, Stephen, *Citizens in War – And After*, Harrap, 1945

Taylor, Michael and Mondey, David, *Guinness Book of Aircraft Facts and Feats*, Guinness Superlatives Ltd., 1984

Turner, John Frayn, *Famous Flights*, Arthur Barker, 1978

Villard, Henry, *Contact! The Story of the Early Birds*, Arthur Barker, 1968

Welch, Ann, *Happy to Fly*, John Murray, 1983

Wright, Norman, *The Dan Dare Dossier*, Hawk Books, 1990

WORKS OF FICTION

Bolton, John, *Mystery Plane*, Mellifont Press, 1930

Burnham, Margaret, *The Girl Aviators and the Phantom Airship*, M.A. Donahue and Co. (Chicago and New York), 1910
The Girl Aviators on Golden Wings, M.A. Donahue and Co. (Chicago and New York), 1911

Carter, Dorothy, *Flying Dawn*, A. & C. Black, 1936
Wings in Revolt, Lutterworth, 1939
Mistress of the Air, Collins, 1939
Star of the Air, Collins, 1940
Snow-Queen of the Air, Collins, 1940
Sword of the Air, Collins, 1941
Comrades of the Air, Collins, 1942
Marise Flies South, Collins, 1944

Drury, C.M., *Kit Norris, Schoolgirl Pilot*, Juvenile Productions Ltd., 1937

Granger, Heather, 'Noreen, Air Hostess', in *Exciting Stories for Girls*, Hamlyn, 1975

Gerling, Robert J., *She'll Never Get Off the Ground*, Coronet, 1971

Hopcroft, G.E., 'Vera Flies Alone', in *Every Girl's Story Book*, Dean & Son, 1940

Hudson, Harriet, *The Sun in Glory*, Headline, 1991

Jefferis, Barbara, *Solo for Several Players*, J.M. Dent, 1961

Johns, Captain W.E., 'The Case of the Black Gauntlet', in *Biggles of the Special Air Police*, Thames Publishing Co., 1953
Biggles Looks Back, Hodder & Stoughton, 1964
Worrals of the WAAF, Lutterworth, 1941
Worrals Carries On, Lutterworth, 1942
Worrals Flies Again, Hodder & Stoughton, 1942
Worrals on the Warpath, Hodder & Stoughton, 1943
Worrals Goes East, Hodder & Stoughton, 1944
Worrals of the Islands, Hodder & Stoughton, 1945
Worrals in the Wilds, Hodder & Stoughton, 1947
Worrals Down Under, Lutterworth, 1948
Worrals Goes Afoot, Lutterworth, 1949
Worrals in the Wastelands, Lutterworth, 1949
Worrals Investigates, Lutterworth, 1950

Knight, Eric, *This Above All*, Cassell, 1941

Krantz, Judith, *Till We Meet Again*, Bantam, 1988

Le Queux, William, *Beryl of the Biplane*, Pearson, 1917

Macmillan, William, 'Flying for the Mounted', in *Empire Annual for Girls*, 'Every Girl's Paper' Office, 1930

Mansfield, Helen, *Some Women Dream*, Collins, 1990
Marsh, Eileen, *Wings at Midnight*, Sampson Low, 1936
Lorna, Air-Pilot, Sampson Low, 1936
Melbourne, Ida, *The Flying Sisters*, Amalgamated Press, 1935
Moorhead, Else, 'Daughter of the Flying Doctor', in *Adventure Stories for Girls*, Hamlyn, 1975
Peck, Winifred, E., 'Her First Flight', in *Blackie Girls' Annual*, 1930
Stainforth, Diana, *Friends and Other Enemies*, Century Hutchinson, 1989
Wentworth, Sally, *Flying High*, Mills & Boon, 1982
York, Duchess of, *Budgie the Little Helicopter*, Simon & Schuster, 1989
Budgie at Bendick's Point, Simon & Schuster, 1989

PERIODICALS

Alice Ottley School Leaflet, 1960, 1963, 1966, 1968, 1989
Biggles and Co., (privately published), nos. 1–7, 1989–91
Boy's Own Paper, Religious Tract Society, 1915–16
British Women Pilots' Association Gazette (privately published), 1991–2
Bunty Summer Special, D.C. Thomson, 1970
Christmas Pie, Odhams, 1937
Collector's Digest, nos. 532, 533, 534 (privately published), 1991
Debbie, D.C. Thomson, 1978
Eagle, Hulton, 1950–60
New Eagle, IPC, 1989–91
Flyin' Jenny (Omnibus), Arcadia (USA), 1987
Forget-Me-Not, Amalgamated Press, 16 March 1916
Girl, Hulton, 1951–2, 1958–60
Girls' Crystal, Amalgamated Press, 1938–9
Girl's Own Paper, Lutterworth Press, 1933–47
Golden Stories, Amalgamated Press, no. 658, 21 March 1911
Home Companion, Amalgamated Press, 30 May 1942
Ideas, Allied Newspapers, no. 1507, 27 January 1934
June, IPC, 1961–7
Justice League of America, DC Comics Ltd. (USA), 1976, 1978, 1982
Legends of the Stargrazers, Innovation Books (USA), 1989
Mandy, D.C. Thomson, 1969–70
Penny Pictorial, Amalgamated Press, 16 March 1918
Pilot, James Gilbert, November 1984
Popular Flying, Pearson, 1934–7
Princess Tina, Fleetway, 1967–8

Puck, Amalgamated Press, October/November 1923

RAF Journal, no. 12, October 1943; vol. 2, no. 6, July 1944

School-Days, Amalgamated Press, 1928–31

Spellbound, D.C. Thomson, nos. 1–30, 1976–8

The Sphere, Sphere & Tatler, 4 May 1918

The Storyteller, Amalgamated Press, June 1935

The Sunbeam, Newnes & Pearson, 2 December 1922

Super Detective Library (Rick Random), Amalgamated Press, no. 111, 3 September 1957; no. 115, 5 November 1957; no. 123, 4 March 1958; no. 133, 7 August 1958

Woman's Journal, Amalgamated Press, 1941–2

Woman's Magazine, Lutterworth, 1930–1; 1935–7

Wonderwoman, DC Comics Ltd. (USA), 1976, 1978, 1990, 1991

INDEX